T0135531

Vom Fachbereich Physik
der Technischen Universität Darmstadt

zur Erlangung des Grades
eines Doktors der Naturwissenschaften (Dr. rer. nat.)

genehmigte Dissertation von
Dipl.-Phys. Christian Kellermann
aus Halle/Saale

Darmstadt 2012

D17

Referent: Prof. Dr. Christian Fischer
Koreferent: PD Dr. Lorenz von Smekal

Tag der Einreichung: 17.10.2011
Tag der Prüfung: 22.02.2012

Bibliografische Information der Deutschen Nationalbibliothek

Die Deutsche Nationalbibliothek verzeichnet diese Publikation in der Deutschen Nationalbibliografie; detaillierte bibliografische Daten sind im Internet über http://dnb.d-nb.de abrufbar.

ISBN 978-3-8325-3156-0

Logos Verlag Berlin GmbH
Comeniushof, Gubener Str. 47,
10243 Berlin
Tel.: +49 (0)30 42 85 10 90
Fax: +49 (0)30 42 85 10 92
INTERNET: http://www.logos-verlag.de

Für Mailin

Zusammenfassung

Das Ziel dieser Dissertation ist die Untersuchung des Massenspektrums von Glueballen in QCD vermittels der Bethe-Salpeter Gleichung unter Benutzung nichtperturbativer Dressingfunktionen für Geist- und Gluonfelder, die aus den zugehörigen Schwinger-Dyson Gleichungen gewonnen werden. Da Glueballe als gebundene Zustände von (auf der Ebene der Lagrangedichte) masselosen Teilchen, welche einer starken Wechselwirkung unterworfen sind, aufgefasst werden, ist für ihre Beschreibung eine Methode notwendig, die sowohl kovariant als auch nichtperturbativ ist.

Im Rahmen der Kontinuumsfeldtheorie ist ein kombinierter Ansatz aus Bethe-Salpeter- und Schwinger-Dyson Gleichungen besonders geeignet.

Im Laufe der Arbeit zeigten sich einige grundlegende Probleme, die gelöst werden mussten.

Zunächst muß eine Bethe-Salpeter Gleichung aufgestellt werden, die für die Untersuchung von Glueballen geeignet ist. Da in dieser Arbeit die Landau-Eichung gewählt wird, treten neben den Gluonen auch Geist-Felder auf, die zu den physikalischen gebundenen Zuständen beitragen. Diese Geist-Felder sind Freiheitsgrade der Gluonen, die nach Eichfixierung als eigenständige Felder erscheinen. Daher müssen bei der Behandlung von Glueballen beide Teilchentypen berücksichtigt werden, deren gebundene Zustände (ob für sich genommen physikalisch oder nicht) miteinander mischen. Es zeigt sich, daß Glueballe (in Landau-Eichung) durch ein System zweier gekoppelter Gleichungen des Bethe-Salpeter-Typs beschrieben werden. Dieses System wird rigoros hergeleitet. Dabei erhalten wir eine generelle Formulierung von Systemen gekoppelter Bethe-Salpeter Gleichungen, die gebundene Zustände beschreiben, welche mischen.

Zweitens werden die expliziten Ausdrücke für die relativistischen Bethe-Salpeter Amplituden hergeleitet, die in der/den Bethe-Salpeter Gleichung(en) benötigt werden. In dieser Arbeit wird dies in sehr allgemeiner Form getan. Wir geben die Amplituden sowohl für beliebigen ganzzahligen Gesamtspin des gebundenen Zustands, als auch für eine große Klasse von verschiedenen Darstellungen der Konstituentenfelder einschließlich Quarks (für Mesonen), Geister und Gluonen (für Glueballe) an.

Als drittes muß das gekoppelte System der Geist- und Gluon Schwinger-Dyson Gleichungen für komplexe Impulse gelöst werden, da die Dressingfunktionen, welche man so erhält, für die konsistente Behandlung der Glueball-Bethe-Salpeter Gleichung unverzichtbar sind. In dieser Arbeit wird dies ebenfalls behandelt. Wir stellen hier Ergebnisse vor und diskutieren diese detailliert.
Schließlich werden wir eine erste Berechnung von Massen von Glueballen im Rahmen des vorgestellten Formalismus durchführen, bei dem wir ein sehr einfaches Trunkierungsschema verwenden, welches konstistent bezüglich der Bethe-Salpeter und der Schwinger-Dyson Gleichungen ist und gegebene Symmetrien erhält. Die Ergebnisse vergleichen wir mit der Literatur, diskutieren sie und geben zuguterletzt einen Ausblick auf Möglichkeiten weiterer Anwendungen.

Abstract

The goal of this thesis is to investigate the mass spectrum of glueballs via the Bethe-Salpeter equation using non-perturbative dressing functions for ghost and gluon fields obtained from their corresponding Schwinger-Dyson equations in QCD. Being a bound state of (on the level of the Lagrangian) massless particles, which are subjected to a strong interaction, glueballs require a treatment which is covariant and non-perturbative. In continuum gauge theory the combined Bethe-Salpeter/Schwinger-Dyson approach is suited particularly well for such a task.

It turns out that, in order to achieve that goal, several fundamental issues have to be dealt with.

The first is to formulate a Bethe-Salpeter equation suited for that problem. Since in this work we use Landau gauge gluons themselves will contribute to the physical bound state, but also ghost fields, which also are degrees of freedom of the gluons, made explicit fields by gauge fixing. Thus there are two fields that have to be taken into account, whose bound states (whether or not they are physical on their own) mix. The resulting bound state (in Landau gauge) is found to be described by a set of two coupled equations of Bethe-Salpeter type. This formalism is derived rigorously. Thus we find a general formulation of systems of coupled Bethe-Salpeter equations for bound states that mix.

Secondly the explicit expressions for the relativistic Bethe-Salpeter amplitudes, which have to be provided as input for the Bethe-Salpeter equation(s), are derived. In this thesis the derivation of Bethe-Salpeter amplitudes is done in a very general way. We give the amplitudes for any integer total spin of the bound state and for a wide range of different representations of the fields of constituent particles, including quarks (suited for mesons), ghosts and gluons (suited for glueballs).

Thirdly the coupled system of Schwinger-Dyson equations for ghosts and gluons has to be solved in the complex momentum plane, since the resulting dressing functions are a vital ingredient for the glueball Bethe-Salpeter equations. In this work we will present and discuss results in detail.

Finally we will do a first calculation of masses of glueballs in the framework developed in this thesis using a very simple truncation scheme, which is consistent with respect to the

Bethe-Salpeter and the Schwinger-Dyson equations and preserves symmetries as well. We will compare our findings to the literature, discuss them and finally give an outlook on possible future applications.

Notes added for the printed version

For the printed version some minor changes have been applied to the original version of this thesis. All graphs have been slightly rescaled for better readability and scalable fonts have been used. Furthermore the plots for the results of ghost and gluon dressing functions in the complex plane have been slightly changed in perspective and different fonts have been introduced. Also for better readability a colour encoding of the function-values has been added. Besides that a number of typos has been corrected.

Contents

List of Figures

List of Tables

Chapter 1

Introduction

The description of hadrons is a long standing problem and challenge of physics. Being compound states bound by the strong interaction they have to be described in terms of Quantum-Chromo-Dynamics (QCD). The simplest such systems are bound states of two quarks or gluons. Bound states of quarks are well known experimentally, for instance nucleons or mesons and several properties of them can be described nicely using QCD e.g. masses and electromagnetic form factors. Bound states of gluons however still are a puzzle, since although they are a key feature of QCD and the necessity of their existence is undisputed, there is no confirmed evidence of an experimental observation of such a gluonic bound state. The problem is that the observable properties of gluonic bound states are a set of quantum numbers and their decay channels, but the quantum numbers can also be found in bound states of quarks namely mesons. Thus only the decay channels can distinguish gluonic bound states from mesons. Therefore very precise inclusive measurements have to be conducted. In order to aim the experimental searches at energies, where bound states of gluons can be found, theoretical investigations should prepare the ground of such searches. This thesis is aimed to contribute to such theoretical studies. In order to elucidate the whole theme we will start with a short historical summary of the emergence of QCD as commonly accepted theory of the strong interaction and then go into details of the general problem of theoretical treatment of QCD with special emphasize on the description of strong bound states.

After the discovery of atomic nuclei by Rutherford, Geiger and Marsden in 1909 [GM09, Rut11], the subsequent discovery of the proton in 1919 [Rut19] and of the neutron in 1932 by Chadwick [CB21], it seemed that a rather complete picture of the existing particles had been found. However with improving experimental technologies plenty of new particles were discovered leading to the notorious "particle-zoo".[1] In the 1950's however the experimental results pointed in the direction that all hadrons are equally elementary. However hard protons were smashed into each other, the fragments observed were hadrons. Furthermore at this time the very concept of quantum field theory was virtually dead. It was considered useful for the case of electrodynamics, since perturbation theory provided the Feynman rules, which are very utile for practical calculations. But the technique of renormalisation was widely considered no more than a calculational trick. Even in 1961 Feynman stated at the Solvay conference "I still hold to this belief and do not subscribe to the philosophy of renormalization." [Sto64].
In the soviet union quantum field theory was abandoned completely after Landau pointed out the zero-charge problem of QED and concluded that the theory is inconsistent [LP55, Pau55].
The methods that were considered most promising for a description of strong interactions at that time were the bootstrap and S-matrix theory [Che61]. The bootstrap was based on the idea that there are no local fields at all and the only physically important entity was the S-matrix, which was to be constructed obeying certain principles like unitarity and analyticity. Furthermore it was assumed that there exist no fundamental particles at all. All the observed particles were assumed to be dynamic bound states of each other. That second principle was called *nuclear democracy*. Having no experimental sign whatsoever of any underlying fundamental degrees of freedom comprising hadrons and a fashion of not even considering microscopic dynamics in the theoretical community, evidently at that time there was very little effort spent on trying to describe the strong interaction in terms of fundamental degrees of freedom. Even in the early 70's talking about quarks was improper without apologising.

[1] The following historical overview is oriented on the excellent outline given by D.J. Gross in his Nobel Lecture [Gro05].

However the symmetries of the strong interaction were under investigation. Most notably the work of Gell-Mann and Ne'eman, who discovered the approximate $SU(3)$ symmetry of strong interactions [GMN64] and of Nambu and Jona-Lasinio who laid the foundation of the understanding of dynamical chiral symmetry breaking [NJL61]. Based on these symmetry considerations Gell-Mann[2] [GM64] as well as Zweig [Zwe64] proposed a hadron model based on fundamental degrees of freedom (certainly not without stressing that these are purely hypothetical objects). The approximate $SU(3)$ symmetry (flavour) was rather accidental at that point, since the quarks comprising the hadrons seen so far in experiments are light. However there remained a problem concerning quantum statistics of baryons. Being considered as a state of three quarks of the same flavour, same spin polarisation and with positive parity the Δ^{++} baryon would be described by a totally symmetric wave function. Having spin 3/2 it is antisymmetric instead. In todays textbooks this is the paradigmatic case, where colour is introduced providing an additional antisymmetric component to the wave-function yielding the correct statistics. In the 1960's the situation was rather nebulous. Greenberg suggested quarks being *parafermions* of degree $p = 3$ [Gre64]. That means that the respective quantum statistic allows for up to three parafermions to be in the same state, leaving states with more than three particles in the same state subject to the Pauli-principle. With this idea it was possible to construct the baryon multiplets with correct statistics. Furthermore it was shown earlier (but published later!) by Greenberg and Messiah that particles obeying parastatistics could not have been observed in experiments on energy levels accessible that time. From todays perspective it is of course immediately clear that this parastatistical behaviour of degree $p = 3$ exactly resembles the colour quantum number. Also Nambu proposed another quantum number a little later to provide an explanation why the interaction assumed to act between quarks does not show up between hadrons[3] [KePeSe65, vHedSeFe65] and together with Han [HN65], he tried to provide a quark model having integer electric charge [CG68]. In 1968 Callan and Gross from a current algebra approach to strong interactions, derived a sum rule which allowed to extract proton structure functions from deep inelastic electron scattering. This was the first

[2] In this paper the notion of "quarks" appears for the first time.

[3] In modern terms: why hadrons are colour singlets.

connection of a model based on the quark picture to experimental observation. From the same sum rule and from dimensional arguments Bjorken suggested that the ratio of electric and magnetic proton structure functions obeys a scaling relation (Bjorken scaling), which was confirmed by experiment soon after. The discovery of Bjorken scaling in deep inelastic scattering together with sum rules derived from current algebra approaches allowed to determine the spin of quarks [CG69] as well as their baryon number [GLS69]. Quarks began to look like something real, being point-like particles with spin 1/2. At the same time Feynman introduced his parton picture [CG69]. However from the work of Callan and Symansik [Cal70, Sym70] it became clear that the observed scaling also means that the scaling behaviour as well as the sum rules by Callan and Gross are only valid in a free theory. But in hadrons the opposite is true the interaction is strong. The breakthrough of the idea of describing the strong interaction by a quantum field theory was achieved in the early 70's when Gross and Wilczek [GW73b, GW73a, GW74] discovered asymptotic freedom in non-abelian gauge theories. In such non-abelian field theories interaction bosons are themselves charged (in contrast to photons in QED which is abelian) and interact among each other. In a relativistic theory this leads to magnetic anti-screening which overcomes screening of electric charges by vacuum-polarisation for short distances if the number of charged fermions in the theory is not too large (in QCD that means smaller than 17 flavours). But an asymptotically free theory could explain Bjorken scaling.

At the same time Fritzsch, Gell-Mann and Leutwyler[4] [FGML73] suggested a gauge theory with quarks and exchange vector-bosons that are colour-octets described by Yang-Mills fields transforming under the adjoint representation of $SU(3)$. This theory was later called Quantum-Chromo-Dynamics (QCD) by Marciano and Pagels [MP78]. Due to asymptotic freedom perturbation theory is applicable in the high-momentum regime and thus it was possible to derive predictions from QCD that could be tested in experiment to high precision. The discovery of additional flavours and quark and gluon jets were milestones which lead to todays believe that strong interaction is indeed described by QCD.

[4] In this paper the notion of colour was introduced the first time.

However even though the symmetries observed in hadron spectra are reproduced by QCD the description of hadronic spectra still faces problems. This is due to the fact that the momentum scales where hadron formation takes place are much lower than those accessible by perturbation theory. For the description of hadronic bound states the application of non-perturbative methods is essential. The probably most successful approach is lattice gauge theory. There are several collaborations which by today have achieved a remarkably good agreement to the observed mass spectrum of mesons and baryons [DFF+08, A+09, BTB+10]. But there are several issues still lacking an explanation like confinement or the nucleon spin puzzle.

The nucleon spin puzzle is a problem of QCD for over 20 years. In the constituent quark model of nucleons the total spin of a nucleon in the ground state is expected to be a superposition of the constituent quarks' spins. However high precision measurements of nucleon structure functions that depend on constituent quark spins [A+88] showed that these quarks contribute only a fraction to the nucleon's spin. From today perspective it is evident that this has to be attributed to the complex structure of strong interaction between the constituent quarks, which are only a good approximation to the nucleon structure at high energies. However at energies where the physics of strong interaction binding takes place, the situation is far more intricate. Although the general reason for the spin puzzle seems somewhat clear, a reasonable description in terms of QCD of this structure of strong interactions at nucleon binding energy scales has not yet been achieved and it seems that there is still a lot of work to be done.

Another problem which this work will focus on are Glueballs. Due to the fact that gluons are colour-charged particles, QCD predicts that there should be observable states consisting solely of gluons, called Glueballs. In fact there was plentiful theoretical work done in a variety of approaches and all agree in the prediction of these states, the probably most precise coming from pure lattice gauge theory. However an experimental confirmation is still missing. Although there have been lots of experiments investigating the hadron mass spectrum (several early experiments for mesons and baryons with light quarks and more recently the CRYSTAL BARREL experiment, charmonia at BNL and the SPEAR storage ring at SLAC, bottomonia in BABAR, BELLE, CLEO and BES experiments and many more for baryons and exotics) it was only possible to find more or less trustworthy glueball candidate states which usually are disputed. The reason for that is that experimentally glueballs can hardly be distinguished from "quarkonic" states by their quantum numbers, since having the same quantum numbers gluonic and quarkonic states mix. Also almost all predictions of glueball masses are done for pure gauge theory. Those which are not, only take into account unquenching effects, neglect final-state mixing and have to rely on staggered quarks [RIGM10].

From the experimental side as mentioned before there are no states that can be deemed more than a candidate for a glueball. Even though there are states which are considered exotic when compared to nonrelativistic quark model predictions, those states could also be tetraquarks, pentaquarks, meson-molecules or something else. Because of these problems, when looking for possible glueball states usually one searches for so-called "gluon-rich" processes, which means that such processes are dominated by intermediate gluon-exchanges. The most prominent examples are radiative J/ψ-decays. In such a reaction a photon and two gluons are created. If the single photon is measured the two remaining gluons must form a glueball or at least decay into something hadronic which would presumably show a resonance at a glueball mass.

Another type of process considered gluon-rich is pomeron-scattering or central-production. That means a reaction where two hadrons pass by each other without direct interaction and thus undergo a diffractive scattering in forward direction. Since no valence-quarks are exchanged in such a reaction, it is assumed that the interaction is purely in terms of gluons.

The third type of reaction is $p\bar{p}$-annihilation. Here intermediate gluons can be vastly created and form glueball states, which decay themselves into hadronic particles. Therefore glueball resonances could well show up in such processes. There has been a lot of work dedicated to the hunt for glueballs and a lot of literature has been written. There are two rather recent reviews on mesons, glueballs, hybrids and other exotics which cover a very broad extent of the topic [KZ07, CM09].

The outstanding problem of finding glueballs is an important part of the research program of two major experimental facilities being under construction today, namely the FAIR facility at GSI in Darmstadt (especially the PANDA experiment) [Joh10, Lan09, Pet07] and the GlueX-experiment at Jefferson-Lab in Newport News, VA [Szc05, She09, Zih10].

Especially due the effects of unquenching and mixing with mesons, it is necessary to provide better theoretical descriptions of glueballs. It is the goal of this work to lay the foundations of a novel approach to the calculation of the glueball mass spectrum, which allows to include unquenching effects, thus providing physical mass spectra of glueballs. The framework adopted in this thesis will be the combined application of Schwinger-Dyson equations (SDE) and Bethe-Salpeter equations (BSE). Both, coming from continuum quantum field theory provide a means of investigating strong interactions complementary to lattice field theory. The advantages of the continuum approach are on the one hand that it is possible to include dynamical quark effects and mixing of final hadronic states in a rather direct way and on the other hand the absence of the notorious problem fermion pose in lattice calculations. Furthermore numerical calculation in the continuum framework are less computational intensive than lattice simulations tend to be. Also in this case the physical mechanism behind the spectrum can possibly be somewhat clearer, since lattice calculations (which are usually done without gauge-fixing) do not resolve the dynamics of the gluon content of the final state they observe.
The latter point should not be taken without a grain of salt however, since the more *physical* picture continuum calculations may provide, is inevitably connected to the most important disadvantage of these calculation, namely being not as *ab initio* as lattice calculations are. The reason for that is, that continuum calculations today always rely on certain approximations. QCD itself is a formally closed and well-defined framework.[5] However in QCD, like in most quantum field theories, it is impossible to find closed non-trivial solutions in general and especially beyond the high-energy limit, where perturbation theory is applicable. In the continuum one can choose to use effective field

[5] At least as well-defined as a quantum-field theory can be. There are many fundamental mathematical subtleties, e.g. that quantum field theories usually are written down in terms of free fields, which are not proper degrees of freedom for an interacting theory. That leads to the problem that the corresponding Hilbert state-spaces of field operators are not connected by a unitary mapping to corresponding Hilbert spaces of the interacting theory. This means that strictly speaking one cannot write down the interacting theory at all, as long as the interacting fields are not known already. Nevertheless doing so in terms of free fields connected with a minimal substitution to include one or more interaction terms, like it is done in the Lagrangian framework, tacitly assumes that there is such a unitary mapping. This being untrue is the deep reason for the emergence of ultraviolet divergences in quantum-field theories and the necessity of renormalisation.

theories, where essentially the theory itself is an approximation to QCD, resembling some set of properties expected to be the most important under the circumstances under consideration. The other possibility is to choose equations from QCD itself and simplify those, which usually means to truncate them in a certain way.
In the last decades there has been a lot of effort spent to develop useful truncation schemes to the SDEs this work will focus on, while the BSE is already a well established means of describing relativistic bound states in nuclear physics as well as in particle physics.

This work will outline the basic ideas of the approach and give an extensive review of the theoretical foundations it is build on as well as present newly developed numerical methods and results that will be compared with the results obtained from lattice calculations and other approaches.
The first part of the thesis will cover the theoretical foundations in detail. It will give a short introduction to QCD, Schwinger-Dyson equations, and the 2PI-effective action formalism. Then a derivation of the Bethe-Salpeter equation in the ladder truncation for fermions will be given along with the generalisation to system of coupled particles. Furthermore it will contain a detailed treatment of formulation of Bethe-Salpeter amplitudes for relativistic bound states of arbitrary spin, parity and charge-conjugation.
In the second part the numerical methods employed to solve the Schwinger-Dyson equations for real and complex momenta and the Bethe-Salpeter equation are detailed. We will apply these to glueballs in pure gauge theory and compare our results against those obtained using lattice methods, Hamiltonian frameworks and constituent gluon models.

In the next chapter will we start our presentation with an introduction to the basic concepts of QCD.

Chapter 2

Basic concepts

2.1 QCD and strong interaction

Today it is commonly accepted that QCD is the theory that describes strong interactions. This is due to several reasons.

- **high energy regime:** The agreement between predictions made from perturbative QCD and high energy experiments is remarkable. Due to asymptotic freedom perturbative QCD provides predictions for a vast variety of cross sections in high energy processes like e^+e^--annihilation, deep inelastic scattering and Drell-Yan production.

- **low energy regime:** Although a rigorous proof is still missing, calculations in various non-perturbative approaches show that QCD includes the crucial phenomena confinement and dynamical chiral symmetry breaking.

- **hadron spectra:** In terms of the constituent quark model observed hadron multiplets can be inferred quite accurately from the QCD flavour structure in a simple way.

QCD is a gauge theory with gauge group $SU(N_C)$[1] (colour). In d dimensions the classical Lagrangian[2] reads[3]

$$\mathcal{L} = \sum_{f=1}^{N_f} \sum_{a,b=1}^{N_C} \sum_{\alpha,\beta=1}^{4} \sum_{\mu,\nu=0}^{d-1} \bar{\psi}_{f,\beta,b} \left(i\, g_{\mu\nu} [\gamma^{\mu}]_{\beta\alpha}\, D^{\nu}_{ba}[A] - m_f\, \delta_{\beta\alpha}\, \delta_{ba} \right) \psi_{f,\alpha,a} \tag{2.1.1}$$

$$- \frac{1}{4} \sum_{\mu,\nu,\rho,\sigma=0}^{d-1} \sum_{a=1}^{N_C^2-1} g_{\mu\rho}\, g_{\nu\sigma}\, F^{\mu\nu}_a[A] F^{\rho\sigma}_a[A], \tag{2.1.2}$$

where

$$F_{\mu\nu,a}[A] = \partial_\mu A_{\nu,a} - \partial_\nu A_{\mu,a} - g \sum_{b,c=1}^{N_C^2-1} f_{abc} A_{\mu,b} A_{\nu,c} \tag{2.1.3}$$

is the non-abelian field strength tensor. We denote with g the *strong* coupling constant, the index f indicates the quark flavours and thus m_f is the bare (unrenormalised) quark mass of the respective flavour. The matrices γ^μ are the usual Dirac matrices, $g_{\mu\nu} = \mathrm{diag}(1, -1, -1, -1)$ is the metric tensor. Furthermore we have the *covariant derivative* acting on a field in the fundamental representation

$$D_{\mu,ab}[A] = \partial_\mu\, \delta_{ab} + i g \sum_{c=1}^{N_C^2-1} A_{\mu c}\, [T^c]_{ab}. \tag{2.1.4}$$

The matrices $[T^c]_{ab}$ are the $N_C \times N_C$-dimensional generators of the *Lie group* $SU(N_C)$ in the fundamental representation. There are $N_C^2 - 1$ such matrices and they satisfy the *Lie algebra*

$$[T^a, T^b] = i \sum_{c}^{N_C^2-1} f^{abc} T_c, \tag{2.1.5}$$

where the f_{abc} are called *structure constants* of the Lie group. The structure constants themselves define another representation of the Lie group by

$$[t_a]_{bc} = -i\, f_{abc} \tag{2.1.6}$$

[1] More precisely for QCD $N_C = 3$, which is what we will use in all numerical calculations. Here we keep N_C for generality.

[2] This general form is often called the Yang-Mills Lagrangian.We work in units where $\hbar = c = 1$.

[3] Throughout this thesis commas or semicolons appearing in indices do *not* denote any kind of derivative.

called the *adjoint representation.* This representation is N_C-dimensional and its elements are $(N_C^2-1)\times(N_C^2-1)$ matrices. The adjoint representation acts on the elements of the Lie group themselves i.e. on the linear space spanned by the generators T^a. The fermion fields $\psi_{f,\alpha,a}$ represent N_f quarks,[4] the vector fields $A_{\mu,a}$ are the gauge bosons called *gluons.* The Lagrangian (2.1.1) is invariant under local $SU(N_C)$ gauge transformations. If we define the gauge transformation operator as

$$[U(\theta(x))]_{ab} = \exp\left(i \sum_{c=1}^{N_C^2-1} \theta_c(x)\,[T_c]_{ab} \right), \tag{2.1.7}$$

the quarks field transform as

$$\psi'_{f,\alpha,b}(x) = \sum_{a,b=1}^{N_C} [U(\theta(x))]_{ab}\,\psi_{f,\alpha,a}. \tag{2.1.8}$$

The gluon field $A_{\mu,a}$ transforms under the adjoint representation. Written out explicitly in the representation space of the adjoint representation the gluon fields are

$$[A_\mu(x)]_{ab} = \sum_{c=1}^{N_C^2-1} A_{\mu,c}(x)\,[T_c]_{ab} \tag{2.1.9}$$

and the gluon fields transform as

$$[A'_\mu]_{ab} = \sum_{c,d=1}^{N_C} [U(\theta(x))]_{ac}\,[A_\mu]_{cd}\,[U^{-1}(\theta(x))]_{db} + \frac{i}{g}\sum_{c=1}^{N_C} (\partial_\mu[U(\theta(x))]_{ac})\,[U^{-1}(\theta(x))]_{cb}. \tag{2.1.10}$$

Plugging (2.1.8) and (2.1.10) into (2.1.1) one can easily check that the QCD Lagrangian is indeed invariant under $SU(N_C)$ gauge transformations.

So far all indices have been written out in detail. In order to simplify the notation in the following the Dirac and flavour indices will in general be suppressed, colour indices will be written explicitly only when necessary. Furthermore the Einstein convention is understood for all types of indices.

We have tacitly assumed that the QCD Lagrangian is quantised in a somewhat straightforward way and denoted the classical Lagrangian (2.1.1) as QCD Lagrangian. However

[4] The index f runs over N_f flavours, α is the Dirac and a the colour index.

it turns out that the quantisation cannot be performed directly. This is due to gauge invariance. This is easily seen, when one tries to do a canonical quantisation of (2.1.1). The conjugate momentum to the field A_μ is

$$\Pi^\mu = \frac{\partial \mathcal{L}}{\partial(\partial_0 A_\mu)}. \tag{2.1.11}$$

However because of the antisymmetry of $F_{\mu\nu}$ the Lagrangian (2.1.1) does not depend on the time-derivative of A_0, thus

$$\Pi^0 = 0. \tag{2.1.12}$$

The canonical quantisation rule is

$$[A_\mu(\mathbf{x},t),\, \Pi^\nu(\mathbf{y},t)] = i\delta_\mu{}^\nu\, \delta^3(\mathbf{x}-\mathbf{y}). \tag{2.1.13}$$

But because of (2.1.12) this gives only three equations for the four component vector field A_μ. In fact the fourth component is subject to gauge symmetry and thus fixed in a more subtle way. Thus a direct canonical quantisation of (2.1.1) is not possible.[5] To circumvent this problem it is convenient to change to the path-integral formalism. Here the quantisation is done by introducing a functional integral over all possible field configurations weighted by the action $\mathcal{S}$ of the theory

$$U = \int \mathcal{D}\phi_i\, \exp\left(i\,\mathcal{S}[\phi_i]\right), \tag{2.1.14}$$

where

$$\mathcal{S} = \int d^4x\, \mathcal{L}. \tag{2.1.15}$$

From the path-integral one can infer time-ordered correlation functions of field operators as expectation values of the respective fields. For instance for the two-point function one finds

$$\langle 0|\, T\left\{\phi(x_1)\phi(x_2)\right\} |\, 0\rangle = \frac{1}{U}\int \mathcal{D}\phi\, \phi(x_1)\, \phi(x_2)\, \exp\left(i\,\mathcal{S}[\phi]\right). \tag{2.1.16}$$

[5] It is possible however to quantise (2.1.1) using *BRST-quantisation* [BRS76, Tyu75] or the *Batalin-Vilkovisky* [BV81] formalism.

For a gauge theory this formalism faces the problem that integrating over all possible configurations of the fields also means integrating over an infinite number of field configurations which can be transformed into one another by a gauge transformation. Such configurations are physically equivalent and the path integration yields an infinite constant corresponding to the "volume" of the space of field configurations connected by gauge transformations. To get rid of this obstacle one introduces a gauge-fixing term in (2.1.1). In order not to break Lorentz-invariance this is most conveniently done using the *Faddeev-Popov* [FP67] method, where a representation of unity in the form

$$\mathbb{1} = \int \mathcal{D}A(x)\, \delta\left(f(A_\theta)\right) \det\left(\frac{f(A_\theta)}{\partial\theta}\right), \tag{2.1.17}$$

where the field A_μ has been subject to a gauge transformation θ

$$A_\theta = A_\mu^a + g^{-1}\, \mathcal{D}_\mu^{ac}\, \theta^c(x) \tag{2.1.18}$$

and f is some function to be chosen for convenience. Furthermore we have the covariant derivative acting on a field in the adjoint representation

$$\mathcal{D}_\mu^{ab} = \partial_\mu\, \delta^{ab} + g\, f^{abc}\, A_\mu^c. \tag{2.1.19}$$

If we choose a linear covariant gauge $f\,(A_\theta)$ has the form

$$f\,(A_\theta) = \partial^\mu\, A_{\mu,a}(x) - \omega_a(x). \tag{2.1.20}$$

Setting the $\omega_a(x)$ to be gaussians centered at $x = 0$, one can Integrate out the gauge condition (2.1.20). The result is simply an additional term in the Lagrangian which reads

$$\mathcal{L}_{gf} = -\frac{1}{2\xi}\left(\partial^\mu A_{\mu,a}\right)^2, \tag{2.1.21}$$

where ξ is the width of the gaussians. However for non-abelian gauge theories the functional determinant (2.1.17) depends on the dynamical fields and can not be factored out of the path integral as it is possible for QED. Hence the Faddeev-Popov procedure introduces another term into the Lagrangian. The functional determinant itself can be represented as a path integral over two scalar Grassmann fields called *ghosts* ans *antighosts.* The additional term in the Lagrangian is

$$\mathcal{L}_{ghost} = -\bar{c}^a\left(\partial^\mu\, \mathcal{D}_\mu^{ab}\right) c^b. \tag{2.1.22}$$

From these considerations we find the QCD Lagrangian in linear covariant gauges

$$\mathcal{L}_{QCD} = \mathcal{L} + \mathcal{L}_{gf} + \mathcal{L}_{ghost}. \tag{2.1.23}$$

The gauge-fixing parameter ξ can be chosen arbitrarily. In this work we will solely work in Landau gauge $\xi = 0$. It turns out however that the Faddeev-Popov procedure does not completely fix the gauge. In case of perturbation theory where fluctuations of the gluon fields are small around the free field point $A = 0$, the Lagrangian (2.1.23) works just fine. If the fluctuations become large however there remain field configurations connected by gauge transformations, called *Gribov-copies*[6] [Gri78]. In fact to obtain a gauge-fixing prescription that holds also non-perturbatively is a somewhat subtle matter. Gribov proposed to restrict the integration domain of the functional integral (2.1.14) to a region Ω which obeys [vB97]

$$\Omega := \{A_\mu(x) : \partial^\mu A_\mu(x) = 0,\ \mathbb{D}[A] > 0\}\,, \tag{2.1.24}$$

where we have defined the Faddeev-Popov operator

$$\mathbb{D}[A] = -\partial^\mu \mathcal{D}^{ab}_\mu. \tag{2.1.25}$$

The domain Ω is called the *first Gribov region.* Its boundary obeys the condition

$$\partial\Omega := \{A_\mu(x) : \partial^\mu A_\mu(x) = 0,\ \mathbb{D}[A] = 0\}\,. \tag{2.1.26}$$

and is called a *horizon.* There is a unique horizon on which the lowest eigenvalue of $\mathbb{D}[A]$ becomes zero, called *first Gribov horizon.*[7] The first Gribov region is the set of all local minima of the norm of the gauge-field vector potential (2.1.10) and can be shown to be convex and to contain at least one member of each gauge orbit [Zwa82], whereby gauge orbit we mean a subset of all field configurations, whose members are connected by a gauge transformation. On the first Gribov region furthermore the Faddeev-Popov

[6] Mathematically the reason for the existence of Gribov-copies is that in a large class of gauges (e.g. linear covariant) it is not possible to impose a global set of affine coordinates on the manifold of gauge-fields, due to topological properties arising from the non-abelian nature of the theory.

[7] The points on the Gribov horizon, in fact are just the coordinate singularities, that result from the fact that the maps between charts covering the gauge field manifold are not diffeomorphic, due to the non-compact topology of the gauge-field manifold.

operator is positive. However of course containing at least one representative of each gauge orbit does not mean exactly one and so the first Gribov region can still contain Gribov copies. In order to get rid of these one defines the *fundamental modular region* as the global minimum of the norm of (2.1.10) [vB97]. To find such a global minimum can however be very complicated. In the framework of Schwinger-Dyson equations the restriction of the path integral to the first Gribov region has been implemented by adding another term to the Lagrangian, usually called the horizon term and the resulting theory is called Gribov-Zwanziger theory [Zwa89].
There have been detailed studies upon the impact and methodology of gauge fixing for continuum field theory as well as lattice field theory and we would like to refer the reader to [Maa10a, Maa10b]. The most important observation from such studies for our purposes is that this problem of gauge-fixing can show up in continuum field theory calculations at some point, possibly an unexpected one. In the continuum approach chosen in this work, we will find, that there is some freedom in choosing the boundary conditions for the dressing function of the ghost. This freedom arises from an imperfect choice of gauge-fixing [Maa10a]. We will discuss that point more detailed in chapter 4.

2.2 Functional methods

In the previous section the path-integral (2.1.14) was introduced. Correlation functions of the respective theory can be inferred from it like (2.1.16). This however is a rather cumbersome method and usually not even possible to do explicitely. A much more elegant way of calculating correlation functions uses the *generating functional* or *partition function*. It also is a path-integral, but with external sources J_i present[8] [Sch51a]

$$Z[J_i] = \int \mathcal{D}\phi_i \exp\left(i\left(\mathcal{S}[\phi_i] + \int d^4x \sum_i \phi_i J_i\right)\right). \tag{2.2.1}$$

[8] We consider a scalar theory with multiple fields. The generalisation to more complicated situations is straightforward.

From this correlation functions can be inferred taking an appropriate functional derivative and setting the sources to zero

$$\langle 0|\, T\left\{\phi_{k_1}(x_1)\ldots\phi_{k_n}(x_n)\right\}|\, 0\rangle = Z[J_i]^{-1}\left(-i\frac{\delta}{\delta\, J_{k_1}(x_1)}\right)\cdots\left(-i\frac{\delta}{\delta\, J_{k_n}(x_n)}\right) Z[J_i]\Bigg|_{J_i=0} . \tag{2.2.2}$$

The above correlation functions still contain disconnected contributions i.e. amplitudes in which not all local fields interact with each other. These can be eliminated using the *generating functional of connected correlation functions*[9] or shorter the *Schwinger functional*

$$i\mathcal{W}[J_i] = \log\left(Z[J_i]\right). \tag{2.2.3}$$

That this functional generates only connected correlation functions can be seen, by simple calculation. For instance the two-point function is found to be[10]

$$\begin{aligned}\left(-i\frac{\delta}{\delta\, J_1}\right)\left(-i\frac{\delta}{\delta\, J_1}\right)\log\left(Z\right)\Bigg|_{J=0} &= \frac{1}{Z}\frac{\delta^2 Z}{\delta\, J_1\,\delta\, J_2} - \frac{1}{Z^2}\frac{\delta\, Z}{\delta\, J_1}\frac{\delta\, Z}{\delta\, J_2}\\ &= \langle\phi_1\,\phi_2\rangle - \langle\phi_1\rangle\,\langle\phi_2\rangle\end{aligned} \tag{2.2.4}$$

and similar calculations lead to higher connected correlation functions. Another very useful functional is obtained from the Schwinger functional by Legendre transformation[11]

$$\Gamma[\langle\phi_i\rangle] = \mathcal{W}[J_i] - \int d^4x \sum_i \langle\phi_i\rangle\, J_i. \tag{2.2.5}$$

It is called the *effective action*. The reason for that is the following: consider the classical action $\mathcal{S}[\phi_i]$. The equations of motion for the classical fields are given by the Euler-Lagrange equations and hence by the functional derivative of $\mathcal{S}[\phi_i]$, which is required to be stationary

$$\frac{\delta\,\mathcal{S}[\phi_i]}{\delta\,\phi_j} = 0. \tag{2.2.6}$$

[9] A proof that $\mathcal{W}$ is indeed the generating functional of connected Green's functions can be found in [Nai05].

[10] We use a shorthand notation for brevity here: $J_1 = J(x_1)$ and $\langle 0|\, T\left\{\phi(x_1)\right\}|\, 0\rangle = \langle\phi_1\rangle$.

[11] It is assumed here that $\mathcal{W}$ is convex.

If we now look at the variation of the effective action we find

$$\begin{aligned}\frac{\delta\,\Gamma[\langle\phi_i\rangle]}{\delta\,\langle\phi_k\rangle} &= \int d^4x \sum_j \frac{\delta\,\mathcal{W}[J_i]}{\delta\,J_j}\,\frac{\delta\,J_j}{\delta\,\langle\phi_k\rangle} - J_k - \int d^4x \sum_j \langle\phi_j\rangle\,\frac{\delta\,J_j}{\delta\,\langle\phi_k\rangle}\\ &= -\,J_k.\end{aligned} \tag{2.2.7}$$

In the absence of external sources the stationary points of the effective action also give equations of motion of the fields ϕ_i. However due to the Schwinger functional being part of the effective action, quantum corrections are already included in the fields.

$$\left.\frac{\delta\,\Gamma[\langle\phi_i\rangle]}{\delta\,\langle\phi_j\rangle}\right|_{J_i=0} = 0. \tag{2.2.8}$$

So the equations of motion for the quantum fields are obtained from the effective action (after setting the sources to zero) as their classical counterparts are from the classical action. Another very important property of the effective action is that it is the generating functional of connected one-particle irreducible (*1PI*) amplitudes. This means that the correlation functions obtained from the effective action include all corrections coming from higher Feynman-diagrams that become disconnected by cutting any line.[12] There is some ambivalence in terminology in the literature here. Some authors call Γ the generating functional of *1PI* functions and use the name effective action for a quantity that most commonly is called *effective potential*.[13]

The effective action is a very useful tool to study a variety of problems in quantum field theory. However its closed explicit form is usually unknown. Therefore one has to resort to approximation schemes. Most importantly such schemes have to preserve the symmetries of the effective action and thus of the theory under consideration. There are several ways to obtain approximations to the effective action, most prominently the loop-expansion or the $1/N$-expansion for gauge theories or theories with multiple numbers of fermions.

The loop-expansion is interesting here, since from it one obtains an expression for the first-order approximation of the effective action that will be useful for the study of the two-particle irreducible effective action that will be introduced in section 3.1. The derivation

[12] The proof of this statement can be found e.g. in [Wei96], [Nai05] and in a very detailed and rigorous way in the excellent book of Alain Connes and Matilde Marcolli [CM08].

[13] The effective potential is the effective action for the special case that the sources do not depend on the spacetime position and thus are invariant under translations.

starts from the observation that there exists an expansion of the effective action in terms of Feynman diagrams.[14] In principle this point of view turns the argument on its head, since the very fact of being a generating functional for Green's functions means that there is a series expansion in terms of such functions, which is resummed into the effective action.[15] A contribution from a Feynman diagram with L loops will have a factor $\hbar^{L-1}$ attached.[16] Thus an expansion in the number of loops of contributing Feynman diagrams will have the general form

$$\hbar^{-1}\,\Gamma[\langle\phi_i\rangle] = \sum_{L=0}^{\infty} \hbar^{L-1}\,\Gamma^{(L)}[\langle\phi_i\rangle]. \tag{2.2.9}$$

Using (2.2.1), (2.2.3) and (2.2.5) one finds that

$$\begin{aligned}\exp\left(\hbar^{-1}\,\Gamma[\langle\phi_i\rangle]\right) &= \int \mathcal{D}\phi_i \,\exp\left(i\hbar^{-1}\left(\mathcal{S}[\phi_i] + \int d^4x \sum_i J_i\left(\phi_i - \langle\phi_i\rangle\right)\right)\right) \\ &= \int \mathcal{D}\phi_i \,\exp\left(i\hbar^{-1}\left(\mathcal{S}[\phi_i + \langle\phi_i\rangle] + \int d^4x \sum_i J_i\,\phi_i\right)\right) \\ &= \int \mathcal{D}\phi_i \,\exp\left(i\hbar^{-1}\left(\mathcal{S}[\phi_i + \langle\phi_i\rangle] + \int d^4x \sum_i J_i\,\frac{\delta\,\Gamma[\langle\phi_i\rangle]}{\delta\,\langle\phi_i\rangle}\right)\right),\end{aligned} \tag{2.2.10}$$

where we have used (or rather assumed) the invariance of the functional integral measure under translations in the second line. Plugging (2.2.9) into (2.2.10), expanding in the field ϕ and using the two versions of the equations of motion (2.2.6) and (2.2.8) we find

$$\exp\left(\hbar^{-1}\,\Gamma[\langle\phi_i\rangle]\right) = \exp\left(\hbar^{-1}\mathcal{S}[\langle\phi_i\rangle]\right)\left(1 + \frac{1}{2\hbar}\sum_i \int \mathcal{D}\phi_i\,\phi_i\left(\frac{\delta^2\mathcal{S}}{\delta\,\phi_i\,\delta\,\phi_i}[\langle\phi_i\rangle]\right)\phi_i + \dots\right). \tag{2.2.11}$$

[14] A more detailed version of this derivation can be found in [Nai05].

[15] It is remarkable however that there exists an *exact* expansion in terms of Feynman diagrams, since it is well known that e.g. the perturbative expansion in general neither does converge, nor does it recover non-perturbative effects. The effective action however does include all such effects and so does its resummed expansion. But neither does necessarily have a closed form.

[16] The factors $\hbar$ will be kept explicit for the remainder of this section.

Evaluating the gaussian in the second term we find for the first two terms in the expansion

$$\Gamma^{(0)}[\langle\phi_i\rangle] = \mathcal{S}[\langle\phi\rangle] \tag{2.2.12}$$

$$\Gamma^{(1)}[\langle\phi_i\rangle] = \frac{1}{2}\,\mathrm{tr}\,\log\left(\sum_i \frac{\delta^2\mathcal{S}}{\delta\,\phi_i\,\delta\,\phi_i}[\langle\phi\rangle]\right). \tag{2.2.13}$$

The second derivative term in the latter equation is nothing but the classical equation of motion. So for the case of a theory with only one field we find the one-loop expansion of the effective action

$$\Gamma[\langle\phi\rangle] = \mathcal{S}[\langle\phi\rangle] + \frac{1}{2}\mathrm{tr}\,\log\left(G_0^{-1}\right) + \Gamma_1[\langle\phi\rangle]. \tag{2.2.14}$$

As stated earlier we will find the above relation useful, when we study the two-particle effective action in section 3.1. We have added the higher order terms of the loop expansion of the effective action in form of the term $G_1[\langle\phi\rangle]$ in (2.2.14). If we take two functional derivatives of $\Gamma[\langle\phi\rangle]$ with respect to $\langle\phi\rangle$, we find for the full propagator G

$$G^{-1}(x,y) = G_0^{-1}(x,y) + 2\frac{\delta^2\,\Gamma_1[\langle\phi\rangle]}{\delta\,\langle\phi(x)\rangle\;\delta\,\langle\phi(y)\rangle}. \tag{2.2.15}$$

With $\Gamma[\langle\phi\rangle]$ being the generating functional of *1PI* correlation functions, it is evident that the double functional derivative in (2.2.15) must be the sum of all self-energy corrections to the classical propagator, which are *1PI.* So in turn the part $\Gamma_1[\langle\phi\rangle]$ must be the generating functional of the corresponding diagrams i.e. it is the sum of vacuum bubbles, which are themselves *1PI.* Besides using some expansion, one can use the full effective action in a formal way. In the next section we will start from this point of view and derive the full quantum equations of motion of a field theory.

2.3 The Schwinger-Dyson equations

The quantum equations of motion for a field theory are the Schwinger-Dyson equations. They are inhomogeneous integral equations for the correlation functions of the respective field theory. The derivation of these equations can be done as follows.[17] Consider a scalar

[17] A detailed derivation like the following applied to various concrete examples in QED and QCD can be found in [RW94].

field theory with only a single field and some general interaction, which is polynomial in the fields. First note the trivial fact that the derivative of a definite integral vanishes if the integration is conducted with respect to the variable of the derivative. This is also true in functional calculus.[18]

$$\begin{aligned} 0 =& \frac{\delta}{\delta\phi} \int \mathcal{D}\phi\, e^{-i\left(\mathcal{S}[\phi]+\int J\phi\right)} \\ =& \int \mathcal{D}\phi \left(\frac{\delta\mathcal{S}}{\delta\phi}[\phi] + J\right) e^{-i\left(\mathcal{S}[\phi]+\int J\phi\right)}. \end{aligned} \tag{2.3.1}$$

To get the term in brackets out of the functional integral, one formally writes it as an operator acting on the partition function. This is possible, because the action is a polynomial in ϕ, since we required the interaction term to be a polynomial in ϕ and the kinetic term is quadratic in the fields.

$$\begin{aligned} \frac{\delta\mathcal{S}}{\delta\phi}[\phi] =& \frac{\delta\mathcal{S}}{\delta\phi}[i\frac{\delta}{\delta J}] \\ \equiv& \sum_k i^k\, (\delta_\phi\mathcal{S})_k \left(\frac{\delta}{\delta J}\right)^k. \end{aligned} \tag{2.3.2}$$

Using relations (2.3.2), (2.2.7) and (2.2.3) we can write (2.3.1) as

$$\left(-\frac{\delta\Gamma}{\delta\langle\phi\rangle} + \frac{\delta\mathcal{S}}{\delta\phi}\left[i\frac{\delta}{\delta J(x)}\right]\right) e^{\mathcal{W}[J]} = 0. \tag{2.3.3}$$

Here we already have an expression for a functional derivative of the effective action. The next step is to get rid of the exponential of the Schwinger function, so that no path-integral remains. To do this we have to bring the exponential to the left of the operator and multiply both sides with $e^{-\mathcal{W}}$. However note that the powers of fields in the interaction term of $\mathcal{S}$ will entail Leibniz-rules when acting on the exponential. By this a multitude of terms can be generated. For instance consider a term quadratic in the fields. Then we have

$$\begin{aligned} (\delta_\phi\mathcal{S})_2 \frac{\delta}{\delta J(x_2)}\frac{\delta}{\delta J(x_1)} e^{\mathcal{W}} =& (\delta_\phi\mathcal{S})_2 \frac{\delta}{\delta J(x_2)} e^{\mathcal{W}} \frac{\delta}{\delta J(x_1)}\mathcal{W} \\ =& (\delta_\phi\mathcal{S})_2\, e^{\mathcal{W}} \left(\left(\frac{\delta\mathcal{W}}{\delta J(x_1)}\right)\left(\frac{\delta\mathcal{W}}{\delta J(x_2)}\right) + \frac{\delta^2\mathcal{W}}{\delta J(x_1)\,\delta J(x_2)}\right) \end{aligned} \tag{2.3.4}$$

[18] In the following the notation will be a bit sloppy for the sake of brevity. Arguments of functions will only be written out, when necessary or as example.

and for a term cubic in the fields we would find

$$\begin{aligned}
&(\delta_\phi \mathcal{S})_3 \frac{\delta}{\delta J(x_3)} \frac{\delta}{\delta J(x_2)} \frac{\delta}{\delta J(x_1)} e^{\mathcal{W}} = \\
&(\delta_\phi \mathcal{S})_3 \, e^{\mathcal{W}} \left(\left(\frac{\delta \mathcal{W}}{\delta J(x_1)} \right) \left(\frac{\delta \mathcal{W}}{\delta J(x_2)} \right) \left(\frac{\delta \mathcal{W}}{\delta J(x_3)} \right) + \left(\frac{\delta \mathcal{W}}{\delta J(x_1)} \right) \left(\frac{\delta^2 \mathcal{W}}{\delta J(x_2) \, \delta J(x_3)} \right) + \right. \\
&\qquad \left(\frac{\delta \mathcal{W}}{\delta J(x_2)} \right) \left(\frac{\delta^2 \mathcal{W}}{\delta J(x_1) \, \delta J(x_3)} \right) + \left(\frac{\delta \mathcal{W}}{\delta J(x_3)} \right) \left(\frac{\delta^2 \mathcal{W}}{\delta J(x_1) \, \delta J(x_2)} \right) + \\
&\qquad \left. \frac{\delta^3 \mathcal{W}}{\delta J(x_1) \, \delta J(x_2) \, \delta J(x_3)} \right). \qquad (2.3.5)
\end{aligned}$$

Similar terms arise from higher powers of the fields in the interaction term of the Lagrangian as well as for terms that contain powers of different kinds of fields. After removing the functional integral from (2.3.3) as stated above one can derive the Schwinger-Dyson equations by applying the appropriate functional derivatives with respect to the expectation values of the fields and setting these expectation values to zero afterwards. There are two relations, which turn out to be useful. The well-known matrix derivation rule

$$\frac{d}{dx} M = -M^{-1} \left(\frac{dM}{dx} \right) M^{-1} \qquad (2.3.6)$$

and a very important relation between the Schwinger functional and the effective action. Starting from the expectation value of the field derived from the Schwinger functional

$$\langle \phi(x) \rangle = \frac{\delta \mathcal{W}}{\delta J(x)}, \qquad (2.3.7)$$

we apply a chain rule and take the functional derivative with respect to the field at position y,

$$\begin{aligned}
\frac{\langle \phi(x) \rangle}{\langle \phi(y) \rangle} &= \int d^4 z \, \frac{\delta^2 \mathcal{W}}{\delta J(x) \, \delta J(z)} \frac{\delta J(z)}{\delta \langle \phi(y) \rangle} \\
&= \delta(x-y) \qquad (2.3.8)
\end{aligned}$$

and using (2.2.7) we find

$$\int d^4 z \, \frac{\delta^2 \mathcal{W}}{\delta J(x) \, \delta J(z)} \frac{\delta^2 \Gamma}{\delta \langle \phi(x) \rangle \, \delta \langle \phi(y) \rangle} = \delta(x-y). \qquad (2.3.9)$$

This important relation means that

$$\frac{\delta^2 \mathcal{W}}{\delta J^2} = \left(\frac{\delta^2 \Gamma}{\delta \langle \phi \rangle^2} \right)^{-1}. \tag{2.3.10}$$

Using these the Schwinger-Dyson equations can be derived. however this procedure can become somewhat cumbersome.
A much more elegant way of deriving the SDEs has been developed in [AHS09]. Starting from (2.3.3), using the functional relation

$$e^{-\mathcal{W}[J]} \left(\frac{\delta}{\delta J} \right) e^{\mathcal{W}[J]} = \frac{\delta \mathcal{W}[J]}{\delta J} + \frac{\delta}{\delta J}, \tag{2.3.11}$$

multiplying with $\exp(-\mathcal{W}[J])$ from the left and using (2.3.7) one finds

$$\frac{\delta \mathcal{S}}{\delta \phi} \left[i \langle \phi \rangle + i\Delta \frac{\delta}{\delta \langle \phi \rangle} \right] - \frac{\delta \Gamma}{\delta \langle \phi \rangle} = 0, \tag{2.3.12}$$

where we have used

$$\frac{\delta}{\delta J} = \frac{\delta^2 \mathcal{W}}{\delta J^2} \frac{\delta}{\delta \langle \phi \rangle} \equiv \Delta \frac{\delta}{\delta \langle \phi \rangle}, \tag{2.3.13}$$

which entails that Δ is the connected two-point function (2.2.4). From (2.3.12) the DSEs can be derived by consecutive application of appropriate functional derivatives.

In the following we will need the SDEs for quarks, gluons and ghosts. We will only state the equations here. Detailed derivations can be found in [RW94]. The SDEs for gluons and ghosts are a coupled system of equations in the following called the *Yang – Mills System*. In Landau-gauge they read[19]

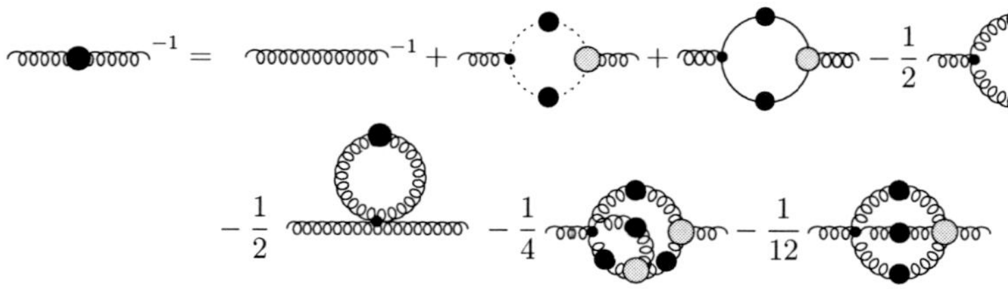

$$[\text{diagrammatic equation}]^{-1} = [\ldots]^{-1} + \ldots + \ldots - \frac{1}{2} \ldots$$

$$- \frac{1}{2} \ldots - \frac{1}{4} \ldots - \frac{1}{12} \ldots \tag{2.3.14}$$

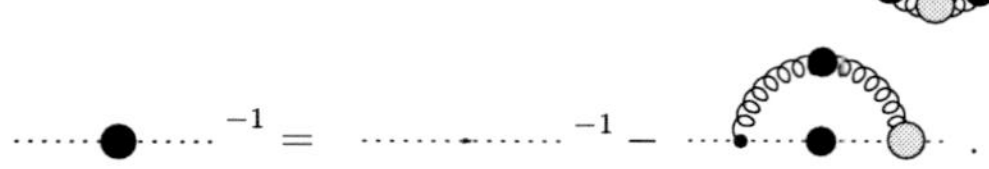

$$[\ldots]^{-1} = [\ldots]^{-1} - \ldots . \tag{2.3.15}$$

[19] We work in euclidean spacetime unless stated otherwise.

The quark SDE is

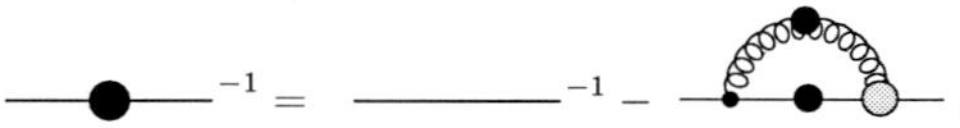

$$\text{[diagram]}^{-1} = \text{[diagram]}^{-1} - \text{[diagram]} . \qquad (2.3.16)$$

Obviously also the quark SDE is coupled to the gluon SDE. Often however one is interested in studying the properties of the gauge-bosons alone and neglects the contribution of quarks and vice versa. Some important properties of the quark SDE do not crucially depend on the precise form of the solution of the gluon SDE. For example dynamical chiral symmetry breaking requires only that the integral over the gluon function is sufficiently large [MR03]. So far most studies of the quark SDE therefore have used model gluons, thus decoupling the quark SDE from the Yang-Mills system [Mar02, MT00, FW08, FW09].
The black blobs in (2.3.14), (2.3.15) and (2.3.16) denote fully dressed propagators, while shaded blobs represent dressed vertices. Working in Landau gauge the propagators have the form

$$\text{[gluon propagator]} = D_{\mu\nu}(p^2) = \left(\delta_{\mu\nu} - \frac{p_\mu p_\nu}{p^2}\right) \frac{Z(p^2)}{p^2} \qquad (2.3.17)$$

$$\text{[ghost propagator]} = D_G(p^2) = -\frac{G(p^2)}{p^2} \qquad (2.3.18)$$

$$\text{[quark propagator]} = S^{-1}(p^2) = i\not{p}\, A(p^2) + B(p^2), \qquad (2.3.19)$$

where $A(p^2)$, $B(p^2)$, $G(p^2)$ and $Z(p^2)$ are scalar dressing functions. The appearance of dressed vertices entails that equations (2.3.14) - (2.3.16) are coupled to SDEs for those vertices. These vertex-SDEs however contain even higher dressed vertices themselves and so forth. That amounts to a coupled system of infinitely many equations. Of course it is not possible to solve such a system and thus an approximation has to be employed. Suitable approximations usually replace the dressed vertices higher than some given order simply by their perturbative version, thus decoupling the system of equations under consideration from the SDEs of higher order. Such an approximation is called a truncation. There are means of refining the approximation scheme at hand. Usually one can include effects of the omitted higher vertices into modifications to dressings of vertices that have been kept. For instance high-momentum behaviour of the solution known from perturbation theory can be used to modify some dressing functions of vertices such, that the solution of the SDEs recovers the perturbative behaviour [DFKS99, DS99, WC04, Fis06]. The details of the truncation scheme that will be employed in this work will be given in chapter 4. Here we give only an preliminary overview.

Considering equations (2.3.14) and (2.3.15) and neglecting quarks, we note that the appearing dressed vertices are the three-gluon vertex, the ghost-gluon vertex and the four-gluon vertex. For technical reasons the diagrams containing dressed four-gluon vertices will be neglected completely. Numerically it is very complicated to calculate multi-loop integrals containing overlapping divergences. Furthermore the knowledge about the four-gluon vertex is still limited. There is no completely self-comsistent investigation of the SDEs up to the four-gluon vertex yet. However a study has been done where an approximation to the (very complicated) four-gluon vertex SDE is investigated [KF08].

In principle the tadpole-diagram will persist, however it can easily be evaluated to give only some constant contribution, which is removed by regularisation. Therefore for the numerical investigation of the Yang-Mills system it also will be dropped. The dressed ghost-gluon vertex is replaced by its bare counterpart. This is a crucial step towards solving the Yang-Mills system and furthermore a quite good approximation especially when one is interested in the low-momentum regime of these equations. The original argument for this approximation was given in [Tay71] and has been shown to be nicely consistent with the result from lattice calculations in [CMM08].

The dressed vertex left is the three-gluon vertex. It will be replaced by its tree-level structure multiplied with a momentum dependent scalar function, which is chosen such that the results for the ghost and gluon dressing functions recover the results obtained in resummed perturbation theory. This leaves some freedom to choose for the general form of the dressing function of the three-gluon vertex and details are given in section 4.1. The resulting system of equations is

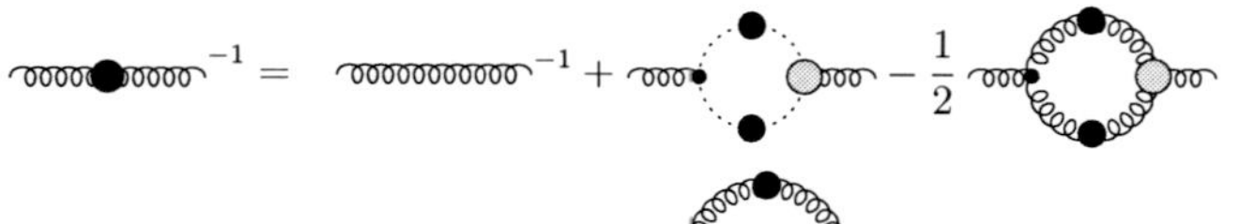

$$^{-1} = {}^{-1} + - \frac{1}{2} \qquad (2.3.20)$$

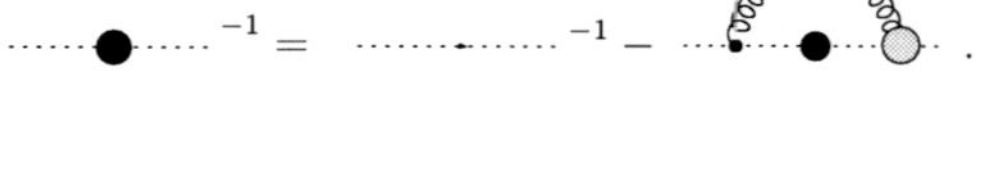

$$^{-1} = {}^{-1} - \; . \qquad (2.3.21)$$

Chapter 3

Advanced concepts

3.1 The 2PI-formalism

In section 2.2 the effective action formalism has been introduced. It allows for the calculation of expectation values of field operators as well as for the derivation of SDEs, the quantum equations of motion. A very important feature in this respect is that the full effective action preserves the symmetries of the respective theory. A possibly serious drawback however is that the closed form of the effective action is usually unknown. One has to resort to approximations like loop or coupling expansions.[1] An alternative closely related is the two-particle irreducible effective action formalism introduced in [CJT74]. It depends not only on source terms coupled to field operators but also on sources coupled to expectation values of two-point correlators including all quantum corrections. It is possible to study not only effective actions like the standard or *1PI* effective action and the *2PI* effective action but also higher versions called *nPI* effective actions,[2] which also depend on proper n-point correlation functions coupled to appropriate source terms. In the absence of sources all these effective actions of course are equivalent. So the different *nPI* effective actions are merely different functional representations of *the* effective

[1] However such expansions usually spoil local symmetries. Global symmetries are still preserved.

[2] A detailed treatment of *nPI* effective actions can be found in [Ber04]. Our following presentation will proceed along the lines given in that paper.

action. However considering a loop expansion one finds that to a given order there is an equivalence hierarchy without vanishing sources. Given an expansion to m-loop order all *nPI* effective action representations with $n \geq m$ are equivalent, while for lower n they are not

$$\Gamma_{1PI}^{(m-loop)} \neq \Gamma_{2PI}^{(m-loop)} \neq \cdots \neq \Gamma_{mPI}^{(m-loop)} = \cdots = \Gamma_{nPI}^{(m-loop)}. \tag{3.1.1}$$

From this one finds that a given loop expansion of an *nPI* effective action resums classes of diagrams appearing in loop expansions for lower effective actions. This allows for systematic approximation schemes to the field theory under consideration, because to a given loop-order the effective action to the same order already represents the complete self-consistent description. It turns out that the loop expansion of an *nPI* effective action is given mainly in terms of a sum of bubble diagrams, which are n-particle irreducible if the background field is zero. Since we will see in the next section, that the kernel of the Bethe-Salpeter equation (BSE) is a sum of two-particle irreducible four-point functions, we will concentrate on the *2PI* effective action in the following.
The standard effective action is given by (2.2.5), which we state again for convenience of the reader

$$\Gamma[\langle\phi_i\rangle] = \mathcal{W}[J_i] - \int d^4x \sum_i \langle\phi_i\rangle \, J_i. \tag{3.1.2}$$

The *2PI* effective action additionally depends on the connected two-point function (2.2.4), which we call G in the following. For brevity we consider some scalar theory with only one type of field. Generalisations are straightforward. We define

$$Z[J,K] = \int \mathcal{D}\phi \, \exp\left(i\left(\mathcal{S}[\phi] + \int d^4x \, \phi \, J + \frac{1}{2}\int d^4x \, d^4y \, \phi \, K \, \phi\right)\right), \tag{3.1.3}$$

where $\mathcal{S}$ is the classical action. As in section 2.2 we have

$$i\mathcal{W}[J,K] = \log\left(Z[J,K]\right). \tag{3.1.4}$$

Obviously

$$\frac{\delta \mathcal{W}[J,K]}{\delta J(x)} = \langle \phi(x) \rangle \tag{3.1.5}$$

$$\frac{\delta \mathcal{W}[J,K]}{\delta K(x,y)} = \frac{1}{2} \left(\langle \phi(x) \rangle \langle \phi(y) \rangle - G(x,y) \right). \tag{3.1.6}$$

Now we perform a double Legendre transformation with respect to J and K

$$\begin{aligned} \Gamma[\langle \phi \rangle, G] = & \mathcal{W}[J,K] - \int d^4x \, \langle \phi(x) \rangle J(x) \\ & - \frac{1}{2} \int d^4x \, d^4y \, \langle \phi(x) \rangle K(x,y) \langle \phi(y) \rangle \\ & - \frac{1}{2} \int d^4x \, d^4y \, G(x,y) K(x,y). \end{aligned} \tag{3.1.7}$$

For vanishing sources we find the stationarity conditions

$$\frac{\delta \Gamma[\langle \phi \rangle, G]}{\delta \langle \phi(x) \rangle} = 0 \tag{3.1.8}$$

$$\frac{\delta \Gamma[\langle \phi \rangle, G]}{\delta G} = 0, \tag{3.1.9}$$

and furthermore we see the relation between *1PI* and *2PI* effective action

$$\Gamma[\langle \phi \rangle] = \Gamma[\langle \phi \rangle, G] \Big|_{G=0} . \tag{3.1.10}$$

The situation for the *2PI* effective action is similar to that in the *1PI* case. We need some systematic approximation scheme to make use of it. The loop expansion is useful in the *2PI* case also. To find the one-loop expansion of the *2PI* effective action we proceed as follows. If we consider the new source term K as a parameter, from (2.2.13) and (3.1.3) we find for the one-loop approximation of the *2PI* effective action

$$\Gamma_K[\langle \phi \rangle] \approx \mathcal{S}_K[\langle \phi \rangle] + \frac{1}{2} \mathrm{Tr} \, \log \left(G^{-1} \right), \tag{3.1.11}$$

where we have defined

$$\mathcal{S}_K[\langle \phi \rangle] = \mathcal{S}[\langle \phi \rangle] + \frac{1}{2} \int d^4x \, d^4y \, \phi \, K \, \phi \tag{3.1.12}$$

$$G^{-1} = G_0^{-1} + K. \tag{3.1.13}$$

Comparing (3.1.7) and (2.2.5) we see that the *2PI* effective action can be written as

$$\Gamma[\langle\phi\rangle, G] = \Gamma_K[\langle\phi\rangle] - \frac{1}{2}\int d^4x\, d^4y\; \langle\phi(x)\rangle\, K(x,y)\, \langle\phi(y)\rangle - \frac{1}{2}\mathrm{Tr}\, KG. \tag{3.1.14}$$

Using (3.1.12) and (3.1.13) the one-loop approximation of the *2PI* effective action reads

$$\Gamma[\langle\phi\rangle, G] \approx \mathcal{S}[\langle\phi\rangle] + \frac{1}{2}\mathrm{Tr}\log G^{-1} + \frac{1}{2}\mathrm{Tr}\, G_0^{-1}G. \tag{3.1.15}$$

Being a series expansion, (3.1.15) can formally be extended to the exact expression for the *2PI* effective action by simply adding a "rest-term". This term is commonly denoted as $\Gamma_2[\langle\phi\rangle, G]$ and the full *2PI* effective action reads [CJT74]

$$\Gamma[\langle\phi\rangle, G] = \mathcal{S}[\langle\phi\rangle] + \frac{1}{2}\mathrm{Tr}\log G^{-1} + \frac{1}{2}\mathrm{Tr}\, G_0^{-1}G + \Gamma_2[\langle\phi\rangle, G]. \tag{3.1.16}$$

To get an understanding what the new term $\Gamma_2[\langle\phi\rangle, G]$ means, we take the variation of (3.1.16) with respect to G and set the result to zero according to (3.1.9). We find

$$G^{-1}(x,y) = G_0^{-1}(x,y) + 2\frac{\delta\,\Gamma_2[\langle\phi\rangle, G]}{\delta\, G(x,y)}, \tag{3.1.17}$$

which looks very similar to (2.2.15). The important difference is that the functional $\Gamma_2[\langle\phi\rangle, G]$ depends on the full propagator already and it is also subject to a functional derivative with respect to the full propagator. It thus depends on a correlation function, into which the *1PI* diagrams already have been resummed. Since it is exact, there cannot be any double counting of diagrams in $\Gamma[\langle\phi\rangle, G]$ and thus the diagrams contributing to the self-energy correction of the propagator in (2.2.15) necessarily have to be of higher irreducibility than *1PI* and thus are *2PI*. So the functional $\Gamma_2[\langle\phi\rangle, G]$ must be the sum of the vacuum bubbles which are *2PI*.

The importance of the *2PI* effective action for our purposes becomes clear, when we remind ourselves that the interaction kernels of the BSE are *2PI* with respect to the constituent particles [SB51, GML51]. Since our goal is a consistent description of relativistic and non-perturbative bound states of particles especially gluons, we need a starting point from which the equations describing the properties of the constituent particles and the equations describing the bound states of those can be derived in a consistent way. Since the SDEs, which are the quantum equations of motion of the constituent particles can be derived from the *2PI* effective action as well as the interaction kernels of the BSE, it is clear that the *2PI* effective action will play an important role in such a derivation. Therefore it is important to find an expression of the *2PI* effective action for QCD, which will automatically produce the desired equations in a certain truncation.

The truncation we choose in this work is in principle already widely determined by the choice of truncation of the SDEs outlined in equations (2.3.20), (2.3.21) and (2.3.16). However there are differences when using the *2PI* effective action.

By construction SDEs depend on both bare and dressed vertices and dressed propagators. In the *2PI* approach self-energy contributions to the propagators which are *1PI* are already resummed from the beginning, so that the propagators appearing in the diagrams are self-consistently dressed ones. This is not the same for the vertex functions. Since the *2PI* effective action does not depend on dressed vertices, in contrast to higher *nPI* effective actions, the vertices appearing in the diagrammatic expressions for the *2PI* effective action are all classical. We thus will only be able to reconstruct equations (3.1.21), (3.1.22) and (3.1.23) with dressed vertices replaced by bare ones.

However to be as realistic as possible, we want the solution of the Yang-Mills system (3.1.21) and (3.1.22) to resemble high-momentum properties known from perturbation theory [vSHA98, FA03]. This can be achieved using appropriate vertex-dressings. Therefore we will have to modify the *2PI* effective action, which we choose to start from, by including model vertex dressings. In order to be consistent we will replace one of the bare vertices by a dressed one, while a suitable symmetrisation is understood and left implicit. This will preserve the symmetries of the *2PI* effective action.

From the effective action we are looking for, we want to generate the following set of equations

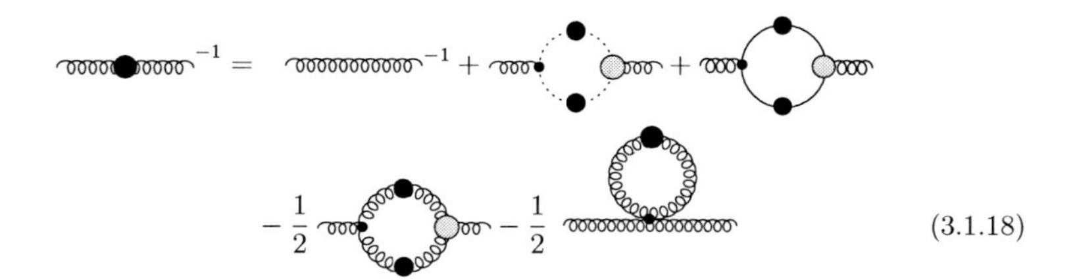

(3.1.18)

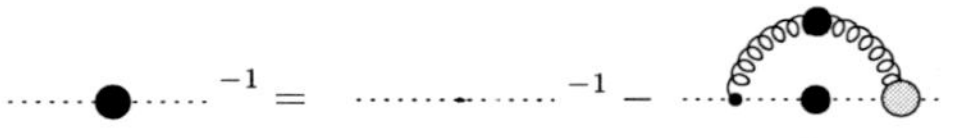

$$\cdots\bullet\cdots^{-1} = \cdots\cdots^{-1} - \tag{3.1.19}$$

$$—\bullet—^{-1} = ——^{-1} - \quad . \tag{3.1.20}$$

The last diagram in (3.1.18) being a tadpole is included for consistency, since it will be generated by the *2PI* effective action and will give rise to an additional contribution to the kernel of the BSE later. In practice this diagram however will be renormalised away. These equations are at most one-loop equations, and it will turn out that the truncation of the BSE following consistently from the corresponding *2PI* effective action will be a kind of generalised ladder-truncation for the Yang-Mills system and the usual ladder-truncation for the meson BSE. To find the expression for the *2PI* effective action of QCD that satisfies our truncation, we start from the stationarity conditions and demand that upon setting all one-point and two-point sources to zero

$$\left.\frac{\delta\,\Gamma_{QCD}}{\delta\,D_{\mu\nu}(x,y)}\right|_{sources\to 0} = 0 \tag{3.1.21}$$

$$\left.\frac{\delta\,\Gamma_{QCD}}{\delta\,G(x,y)}\right|_{sources\to 0} = 0 \tag{3.1.22}$$

$$\left.\frac{\delta\,\Gamma_{QCD}}{\delta\,S(x,y)}\right|_{sources\to 0} = 0, \tag{3.1.23}$$

we find the equations (3.1.18), (3.1.19) and (3.1.20). The one-loop expansion with "rest-term" of the desired effective action reads[3]

$$\begin{aligned}\Gamma_{QCD} =& \frac{1}{2}\mathrm{Tr}\log D^{-1} + \frac{1}{2}\mathrm{Tr}\, D_0^{-1}D - \mathrm{Tr}\log G^{-1} - \mathrm{Tr}\, G_0^{-1}G \\ &- \mathrm{Tr}\log S^{-1} - \mathrm{Tr}\, S_0^{-1}S + \Gamma_{QCD,2}.\end{aligned} \tag{3.1.24}$$

[3] Since the classical action $\mathcal{S}_{QCD}$ does not explicitly depend on the two-point correlators, it will drop out applying any derivative with respect to a propagator to the effective action. We will thus neglect the term from the beginning.

From this, using (3.1.21), (3.1.22) and (3.1.23) we find

$$D_{\mu\nu}^{-1}(x,y) = D_{0,\mu\nu}^{-1}(x,y) + \frac{\delta\,\Gamma_{QCD,2}}{\delta\,D_{\mu\nu}(x,y)} \tag{3.1.25}$$

$$G^{-1}(x,y) = G_0^{-1}(x,y) - \frac{\delta\,\Gamma_{QCD,2}}{\delta\,G(x,y)} \tag{3.1.26}$$

$$S^{-1}(x,y) = S_0^{-1}(x,y) - \frac{\delta\,\Gamma_{QCD,2}}{\delta\,S(x,y)}. \tag{3.1.27}$$

Diagrammatically, taking a functional derivative with respect to a propagator means "cutting" an internal line of the corresponding propagator-type. From the Leibniz rule it is clear, that if there are several possibilities to do so, the result is simply the sum of all possible ways to cut the line. Having this in mind, it is easy to find the desired representation for $\Gamma_{QCD,2}$

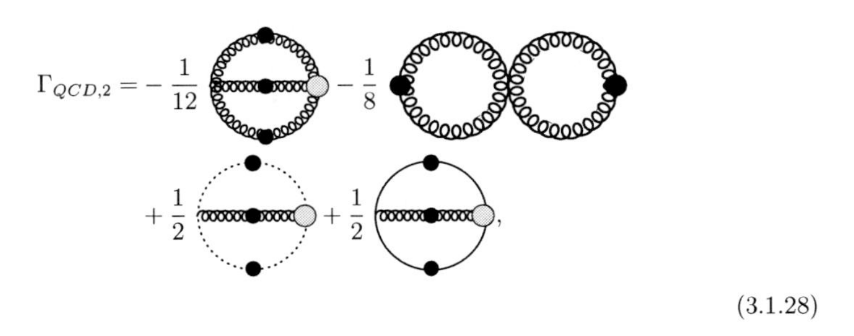

(3.1.28)

which up to the dressed vertices is consistent with the true *2PI* effective action given in [Ber04]. From this we will derive a system of BSEs for glueballs and mesons in section 3.2.2. The deeper details of the truncation scheme will be discussed in chapter 4 in context with the numerical treatment of the equations (3.1.18), (3.1.19) and (3.1.20) and the BSEs.

3.2 The Bethe-Salpeter equation

3.2.1 The standard derivation of the homogeneous BSE

In quantum field theory bound states of two particles are described by the Bethe-Salpeter equation (BSE). It was first written down without derivation by Nambu [Nam50] and shortly afterwards derived by Bethe and Salpeter [SB51] as well as Gell-Mann and Low [GML51]. These derivations however relied on loop-expansions and counting of diagrams, which made them somewhat dependent on the validity of such loop-expansions, which is not necessarily given beyond perturbation theory. The first rigorous derivation was provided by Schwinger [Sch51a, Sch51b], who made use of variational techniques without resorting to any diagrammatic methods. For the purpose of this thesis the original inhomogeneous BSE is not of particular interest on its own, since it is much simpler to generalise the homogeneous version of the BSE, which is what will be important in the following. Therefore the inhomogeneous BSE will be given without derivation or proof and instead the derivation of the homogeneous BSE from it will be reviewed. Then an alternative and direct derivation of the homogeneous BSE will be given and shown to be equivalent to the original. The advantage of the second method is that it is possible to generalise it to situations where bound states of different particles mix in a very natural and straightforward way. It does however use the formalism of two-particle irreducible effective actions. The basics of this concept have been outlined in section 3.1
To start let us consider the inhomogeneous BSE for a fermion-antifermion pair, which is written diagrammatically as[4]

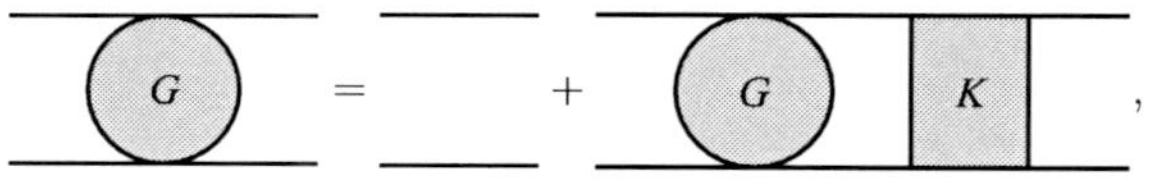

G = + G K , (3.2.1)

where G is the fermion-antifermion Green's function

$$G = \langle 0 | \, T\{\bar{\psi}(x_1)\psi(x_2)\bar{\psi}(y_1)\psi(y_2)\} \, | 0 \rangle \tag{3.2.2}$$

[4] In this section we will work in Minkowski spacetime.

and K is the sum of all diagrams, which are two-particle irreducible (*2PI*) with respect to the constituent fermion lines i.e. which cannot be disconnected by drawing a single line that cuts every constituent fermion line exactly once and no exchange particle lines at all. From this equation the homogeneous BSE can be derived. Originally this was done be Mandelstam [Man55]. A very detailed derivation has been given by Nakanishi [Nak69] and we will mainly follow it here. First note that in order to represent a two-particle bound state, equation (3.2.2) has to have a pole when the total momentum of the state equals the bound state mass [Ede52]. Inserting a complete set of two-particle states $\langle s|$, G reads

$$G(x_1, x_2, y_1, y_2) = \sum_s \langle 0|\, T\{\bar{\psi}(x_1)\psi(x_2)\}\,|\, s\rangle\, \langle s|\, T\{\bar{\psi}(y_1)\psi(y_2)\}\,|\, 0\rangle\,. \tag{3.2.3}$$

We define the Bethe-Salpeter wave-functions to be

$$\chi_s(x_1, x_2; P) = \langle 0|\, T\{\bar{\psi}(x_1)\psi(x_2)\}\,|\, s\rangle \tag{3.2.4}$$

$$\bar{\chi}_s(y_1, y_2; P) = \langle s|\, T\{\bar{\psi}(y_1)\psi(y_2)\}\,|\, 0\rangle\,, \tag{3.2.5}$$

where P is the total momentum of the fermion-antifermion state. Due to Poincaré invariance, the bound state is invariant under translations and we can separate the dependence on the center-of-mass coordinate and write [Man55]

$$\chi_s(x_1, x_2; P) = e^{-iPX}\chi_s(x, P) \tag{3.2.6}$$

$$\bar{\chi}_s(y_1, y_2; P) = e^{iPX}\bar{\chi}_s(y, P), \tag{3.2.7}$$

where x and y on the RHS are the respective relative coordinates. Suppose there is a non-degenerate bound state $|\, s_B\rangle$ in (3.2.3) with bound state mass P_B^2. In some finite momentum region around P_B^2 only that state will contribute to the sum over complete states, since G has a pole at P_B^2. For G we can thus expect the general form

$$\begin{aligned} G(x, y; P) &= \frac{\langle 0|\, T\{\bar{\psi}(x_1)\psi(x_2)\}\,|\, s_B\rangle\, \langle s_B|\, T\{\bar{\psi}(y_1)\psi(y_2)\}\,|\, 0\rangle}{P^2 - P_B^2 + i\epsilon} \\ &\qquad + \text{terms regular in } (P^2 - P_B^2) \\ &= \frac{\chi_B(x, P_B)\bar{\chi}_B(y, P_B)}{P^2 - P_B^2 + i\epsilon} + \text{terms regular in } (P^2 - P_B^2). \end{aligned} \tag{3.2.8}$$

We now switch to momentum space. The inhomogeneous BSE (3.2.1) written out explicitly then reads

$$G(p,q,P) = S(p+\eta\, P)\, S(p-(1-\eta)\, P) + \int d^4p'\, K(p',q,P)\, G(p',q,P), \tag{3.2.9}$$

where η is the parameter that describes the fraction of the total momentum carried by one of the particles. We now insert the Fourier transformed version of (3.2.8) into (3.2.9), multiply both sides with a factor $(P^2 - P_B^2)$ and take the limit $P^2 \to P_B^2$. All regular terms then vanish and we find

$$\chi_B(p,P_B)\bar{\chi}_B(q,P_B) = \int d^4p'\, K(p',q,P)\, \chi_B(p',P_B)\bar{\chi}_B(q,P_B) \tag{3.2.10}$$

and on comparison of both sides of (3.2.10) we find the homogeneous BSE

$$\chi_B(p,P_B) = \int d^4p'\, K(p',q,P_B)\, \chi_B(p',P_B). \tag{3.2.11}$$

The kernel K still is the sum of all diagrams that are *2PI* with respect to the constituents. In the derivation we have neglected several normalisation factors of Fourier-transformations, since they can be absorbed into the definition of $\chi_B(p,P_B)$ in (3.2.11), because it is a homogeneous equation. Obviously $\chi(p,P_B)$ is an eigenfunction to the kernel K, which is sometimes called the Bethe-Salpeter operator. If we had not considered fermions but bosons instead, the only difference would have been the appearance of another inhomogeneous term in (3.2.1) namely crossed propagators [Sch51b]. However this term would also have dropped out like the inhomogeneous term in the fermionic case, so the form of the homogeneous BSE is the same for fermions and bosons.

In the next section we will give another derivation of the homogeneous BSE, starting from a completely different view-point, namely from the *2PI*-property of the kernel K together with the requirement that the one-particle propagator be a stable solution of the quantum-equations of motion.

3.2.2 An alternative derivation of the homogeneous BSE

In this section we will derive the homogeneous BSE again from a different viewpoint than in the previous. The derivation given there is suited for the case of relativistic bound states of two particles which are not coupled to bound states of other particles. In Landau gauge Yang-Mills theory however this is not the case, since here some unphysical degrees of freedom of gluons are accounted for in terms of ghosts. On the other hand there are ghost-free gauges in which glueballs are described entirely in terms of bound states made up of gluons.

However the glueball itself is a physical particle and thus independent of the gauge. Unless there would be some mechanism[5] which removes all contributions from unphysical degrees of freedom of gluons to glueball states, it is necessary to take into account all degrees of freedom of the gluons in any calculation of glueball properties. That unphysical degrees of freedom have no impact at all to glueball properties is highly unlikely,[6] since this is the case not even in simple calculations in perturbation theory. It is therefore necessary to take contributions from ghost degrees of freedom into account explicitly, which means to include mixing of glueball states made up from gluons and "bound states" made up from ghosts. The latter we will call *ghostballs*.

Besides this, the possibility of taking into account several mixing bound states of various particle types is crucial for the description of a realistic glueball and meson spectrum, since meson states and glueballs of course do mix. The formalism outlined in this section is capable of doing this. It allows for a consistent description of these mixing states in terms of coupled systems of homogeneous BSEs. We will give the derivation directly for the cases of mesons and glueballs in QCD. We start with the *2PI* effective action of QCD to two loop order as derived in section 3.1

$$\begin{aligned}\Gamma[D,G,S] =& \frac{1}{2}\operatorname{Tr}\ln D^{-1} + \frac{1}{2}\operatorname{Tr} D_0^{-1}D - \operatorname{Tr}\ln G^{-1} - \operatorname{Tr} G_0^{-1}G \\ & - \operatorname{Tr}\ln S^{-1} - \operatorname{Tr} S_0^{-1}S + \Gamma_2[D,G,S],\end{aligned} \tag{3.2.12}$$

[5] Or rather infinitely many such, since these mechanisms are of course themselves gauge-dependent.

[6] If this was the case however, the formalism still would be correct since such a mechanism would merely decouple the resulting system of equations.

where $\Gamma_2[D,G,S]$ sums up all *2PI* vacuum graphs and is represented graphically up to two loop order as

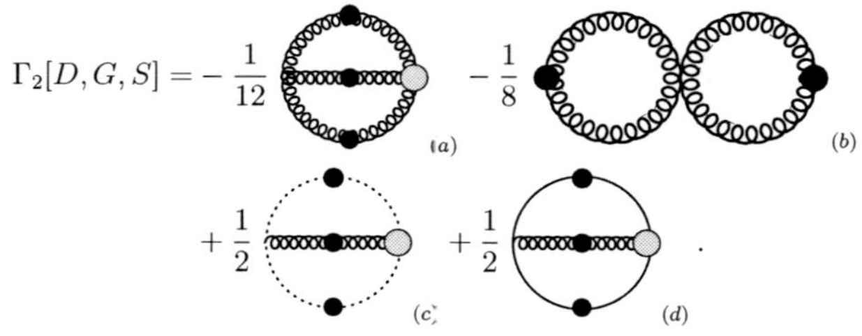

$$\Gamma_2[D,G,S] = -\frac{1}{12}\,(a) - \frac{1}{8}\,(b) + \frac{1}{2}\,(c) + \frac{1}{2}\,(d)\,. \tag{3.2.13}$$

By variation of the effective action with respect to a propagator, one finds the corresponding Schwinger-Dyson equations

$$\frac{\delta\Gamma[D,G,S]}{\delta D} = -\frac{1}{2}D^{-1} + \frac{1}{2}D_0^{-1} + \Sigma_D[D,G,S] = 0 \tag{3.2.14}$$

$$\frac{\delta\Gamma[D,G,S]}{\delta G} = G^{-1} - G_0^{-1} + \Sigma_G[D,G,S] = 0 \tag{3.2.15}$$

$$\frac{\delta\Gamma[D,G,S]}{\delta S} = S^{-1} - S_0^{-1} + \Sigma_S[D,G,S] = 0, \tag{3.2.16}$$

where we have

$$\Sigma_A = \frac{\delta\Gamma_2}{\delta A}.$$

The diagrams which are represented by the self-energies Σ are generated diagrammatically from Γ_2 by cutting one propagator line corresponding to the functional derivative. We now proceed to derive the homogeneous BSE for mesons first, following the work of [7], proving the equivalence of the formalism adopted, to the classical derivation and then apply it to bound states of ghosts and gluons. The idea is to look at the stability condition of a solution to the quantum equations of motion. It will turn out that these are connected to BSE amplitudes. We will first outline the derivation and then show that indeed it leads to the homogeneous BSE.Suppose that $\hat{S}$ is a solution to the equation of motion (3.2.16). If we want to investigate the stability of this solution we add a small

[7] The factors $\hbar$ will be kept explicit for the remainder of this section.

"perturbation" δ_S and expand (3.2.16) to first order keeping only functional derivatives with respect to S. The case where all derivatives are taken under consideration will be reviewed later. We find

$$\begin{aligned}\frac{\delta\Gamma[D,G,S]}{\delta\,S(x,y)}\bigg|_{\hat{S}+\delta_S} &\approx \frac{\delta\Gamma[D,G,S]}{\delta\,S(x,y)}\bigg|_{\hat{S}} + \int d^4x' d^4y' \frac{\delta^2\Gamma[D,G,S]}{\delta\,S(x,y)\delta\,S(x',y')}\bigg|_{\hat{S}} \delta_S(x',y') \\ &\approx \int d^4x' d^4y' \frac{\delta^2\Gamma[D,G,S]}{\delta\,S(x,y)\delta\,S(x',y')}\bigg|_{\hat{S}} \delta_S(x',y'),\end{aligned} \tag{3.2.17}$$

where in the last step we have used the equation of motion (3.2.16). In order for the solution $\hat{S}$ to be stable, it is necessary that (3.2.17) vanishes

$$\int d^4x' d^4y' \frac{\delta^2\Gamma[D,G,S]}{\delta\,S(x,y)\delta\,S(x',y')}\bigg|_{\hat{S}} \delta_S(x',y') = 0. \tag{3.2.18}$$

From the identity

$$\int d^4x'\, d^4y' \frac{\delta^2\,W}{\delta\,J_S(x,y)\,\delta\,J_S(x',y')}\frac{\delta^2\,\Gamma}{\delta\,S(x,y)\,\delta\,S(x',y')} = \delta(x-x')\,\delta(y-y'), \tag{3.2.19}$$

we see that $\delta^2\,\Gamma/(\delta\,S\delta\,S)$ is the inverse of $\delta^2\,W/(\delta\,J_S\delta\,J_S)$ in the functional sense. We furthermore see from (3.2.18) that $\delta_S(x,y)$ is an eigenvector of $\delta^2\,\Gamma/\delta\,S\delta\,S$ also in the functional sense, corresponding to eigenvalue $\lambda = 0$. It is thus also an eigenvector to $\delta^2\,W/\delta\,J_S\delta\,J_S$, with $1/\lambda = 0$. In order for (3.2.19) to hold, $\delta^2\,W/\delta\,J_S\delta\,J_S$ has to have a pole, when $\delta_S(x,y)$ is an eigenvector. So besides (3.2.18) we find

$$\int d^4x' d^4y' \frac{\delta^2\,W[J_D,J_G,J_S]}{\delta\,J_S(x,y)\delta\,J_S(x',y')}\xi(x',y') = \frac{1}{\lambda}\,\xi(x',y'). \tag{3.2.20}$$

at some pole of $\delta^2\,W/\delta\,J_S\delta\,J_S$.

The eigenvector $\xi(x,y)$ is proportional to $\delta_S(x,y)$ and gives the residue of the pole $\delta^2 W/\delta J_S \delta J_S$. Since

$$\frac{\delta^2 W}{\delta J_S(x,y)\, \delta J_S(x',y')} = \langle 0|\, T\{\bar{\psi}(x)\, \psi(y)\, \bar{\psi}(x')\, \psi(y')\}\, |\, 0\rangle \tag{3.2.21}$$

is the connected fermion-antifermion four-point function, we know that the poles of $\delta^2 W/\delta J_S \delta J_S$ correspond to fermion-antifermion bound states. With $\xi(x,y)$ giving the residues at these poles and being eigenvectors of $\delta^2 W/\delta J_S \delta J_S$, by comparison with (3.2.11), we see that indeed the $\xi(x,y)$ are precisely the Bethe-Salpeter amplitudes. We can make this fact manifest by evaluating

$$\begin{aligned} \frac{\delta^2 \Gamma[D,G,S]}{\delta S(x,y) \delta S(x',y')} &= \frac{\delta}{\delta S(x,y)} \left(S^{-1}(x',y') - S_0^{-1}(x',y') - \frac{\delta \Gamma_2[D,G,S]}{\delta S(x',y')} \right) \\ &= -\, S^{-1}(x,x')\, S^{-1}(y,y') - \frac{\delta^2 \Gamma_2[D,G,S]}{\delta S(x,y)\, \delta S(x',y')}. \end{aligned} \tag{3.2.22}$$

For the last step we have chosen to view $S(x,y)$ as a function on some discretised spacetime grid. Then S would be a matrix and we can use the identity $\partial A^{-1}_{ij}/\partial A_{kl} = -A^{-1}_{ik}\, A^{-1}_{jl}$. Taking the limit of vanishing spacing leads to the corresponding identity for the function $S(x,y)$. Plugging (3.2.22) into (3.2.18) and defining

$$\chi(x,y,P) = \int d^4x'\, d^4y'\, S^{-1}(x,x')\, S^{-1}(y,y')\, \delta_S(x',y') \tag{3.2.23}$$

$$K(x,y,x',y') = \int d^4z\, d^4z'\, \frac{-\delta^2\, \Gamma_2[D,G,S]}{\delta S(x,y)\, \delta S(z,z')} S(x',z)\, S(y',z'), \tag{3.2.24}$$

we find

$$\chi(x,y,P) = \int d^4x'\, d^4y'\, K(x,y,x',y')\, \chi(x',y',P), \tag{3.2.25}$$

which is just the position space version of (3.2.11), obtained by going to the center-of-mass frame and a subsequent Fourier transformation of the remaining spacetime variable. Using (3.2.24) it is now easy to determine $K(p,q,P)$ and the BSE of a fermion-antifermion system in the two-loop approximation of the *2PI* effective action. We cut all diagrams in

(3.2.13) twice with respect to a fermion line and find

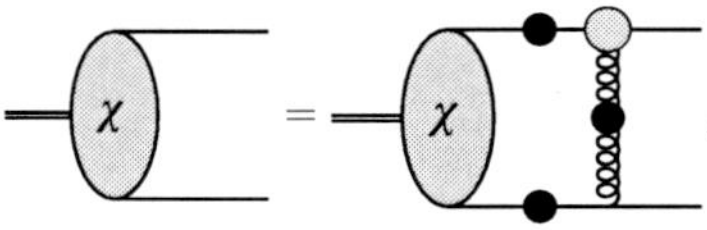

(3.2.26)

which corresponds to the standard rainbow-ladder truncation.

In the same way we can derive BSEs for gluons and ghosts. In (3.2.17) we kept only functional derivatives with respect to fermion lines. To do so in the case of gluons also would be rather questionable since the ghosts are degrees of freedom of the gluons themselves and appear only due to gauge-fixing. Thus ghosts and gluons have to be treated on equal footing and mixed derivatives cannot be neglected.

Consider an effective action of the Yang-Mills sector of QCD in the same approximation as (3.2.13). Obviously that means only omitting the diagram with fermion lines. We proceed analogously to (3.2.17) and assume the $\hat{D}$ and $\hat{G}$ are solutions of the equations of motion (3.2.14) and (3.2.15). In the following we will use a shorthand notation omitting the spacetime arguments and indicating primed arguments by primed functions. This time we expand in two variables and again keep only the linear terms

$$\begin{aligned}\left.\frac{\delta\Gamma[D,G]}{\delta D}\right|_{\hat{D}+\delta_D,\hat{G}+\delta_G} \approx& \left.\frac{\delta\Gamma[D,G]}{\delta D}\right|_{\hat{D},\hat{G}} \\ &+ \int d^4x' d^4y' \left.\frac{\delta^2\Gamma[D,G]}{\delta D\delta D'}\right|_{\hat{D},\hat{G}} \delta'_D \\ &+ \int d^4x' d^4y' \left.\frac{\delta^2\Gamma[D,G]}{\delta D\delta G'}\right|_{\hat{D},\hat{G}} \delta'_G \end{aligned} \tag{3.2.27}$$

$$\begin{aligned}\left.\frac{\delta\Gamma[D,G]}{\delta G}\right|_{\hat{D}+\delta_D,\hat{G}+\delta_G} \approx& \left.\frac{\delta\Gamma[D,G]}{\delta G}\right|_{\hat{D},\hat{G}} \\ &+ \int d^4x' d^4y' \left.\frac{\delta^2\Gamma[D,G]}{\delta G\delta D'}\right|_{\hat{D},\hat{G}} \delta'_D \\ &+ \int d^4x' d^4y' \left.\frac{\delta^2\Gamma[D,G]}{\delta G\delta G'}\right|_{\hat{D},\hat{G}} \delta'_G. \end{aligned} \tag{3.2.28}$$

Using again the equations of motion we require for the solutions $\hat{D}$ and $\hat{G}$ to be stable that

$$\int d^4x' d^4y' \frac{\delta^2\Gamma[D,G]}{\delta D \delta D'}\bigg|_{\hat{D},\hat{G}} \delta'_D + \frac{\delta^2\Gamma[D,G]}{\delta D \delta G'}\bigg|_{\hat{D},\hat{G}} \delta'_G = 0 \tag{3.2.29}$$

$$\int d^4x' d^4y' \frac{\delta^2\Gamma[D,G]}{\delta G \delta D'}\bigg|_{\hat{D},\hat{G}} \delta'_D + \frac{\delta^2\Gamma[D,G]}{\delta G \delta G'}\bigg|_{\hat{D},\hat{G}} \delta'_G = 0. \tag{3.2.30}$$

The functional derivatives of the effective action can be evaluated as

$$\frac{\delta^2\Gamma[D,G]}{\delta D(x,y)\delta D(x',y')} = -\, D^{-1}(x,x')\, D^{-1}(y,y') - \frac{\delta^2\,\Gamma_2[D,G]}{\delta D(x,y)\,\delta D(x',y')} \tag{3.2.31}$$

$$\frac{\delta^2\Gamma[D,G]}{\delta D(x,y)\delta G(x',y')} = -\, \frac{\delta^2\,\Gamma_2[D,G]}{\delta D(x,y)\,\delta G(x',y')} \tag{3.2.32}$$

$$\frac{\delta^2\Gamma[D,G]}{\delta G(x,y)\delta D(x',y')} = -\, \frac{\delta^2\,\Gamma_2[D,G]}{\delta G(x,y)\,\delta D(x',y')} \tag{3.2.33}$$

$$\frac{\delta^2\Gamma[D,G]}{\delta G(x,y)\delta G(x',y')} = -\, G^{-1}(x,x')\, G^{-1}(y,y') - \frac{\delta^2\,\Gamma_2[D,G]}{\delta G(x,y)\,\delta G(x',y')}. \tag{3.2.34}$$

With definitions similar to (3.2.23) and (3.2.24) we find a coupled system of BSEs for ghost and gluon bound states[8]

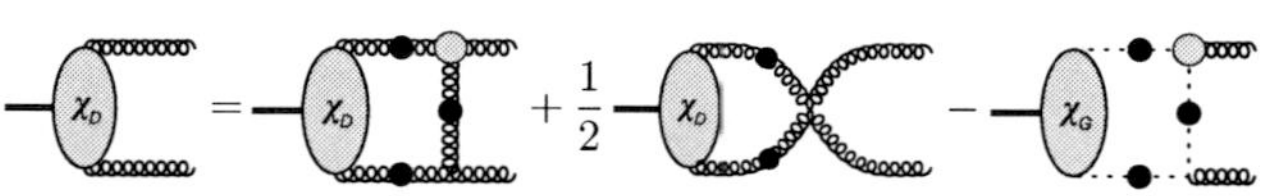

(3.2.35)

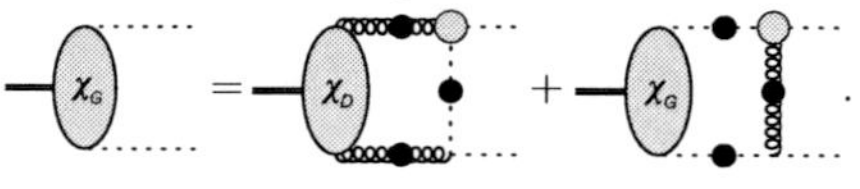

(3.2.36)

Considering the full effective action (3.2.13), we can apply the same derivation and keep functional derivatives with respect to all types of propagators. Evidently we find the full

[8] Note that the vertex blob is a shorthand indicating that the diagram is properly symmetrised.

system of coupled BSEs

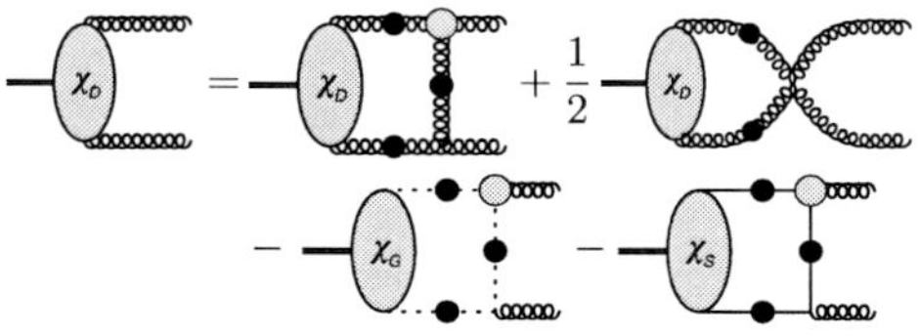

(3.2.37)

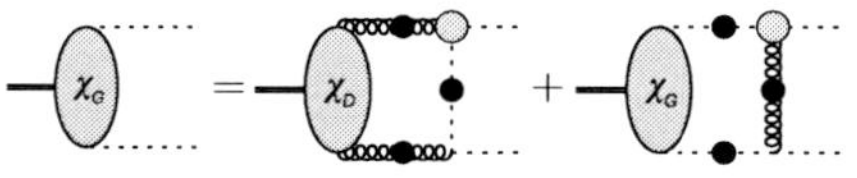

(3.2.38)

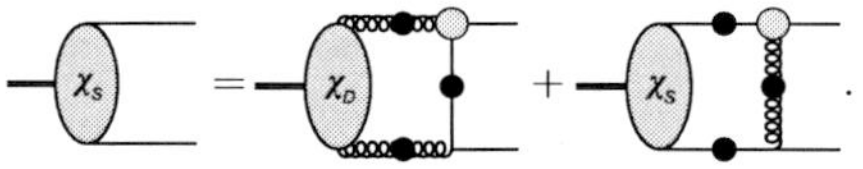

(3.2.39)

This set of equations describes mesons and glueballs in the generalised ladder truncation. Note on the right hand sides the last diagram in (3.2.37) and the first of (3.2.39). They provide for glueball/meson mixing. Especially the latter is interesting. Obviously being present only in the flavour singlet channel[9] from this diagram we expect a contribution to η-η' splitting. In calculations of meson properties using the meson BSE, usually one uses light quarks in the isospin limit and possibly strange quarks, whose masses are fit such that the observed masses of certain, well-known mesons like the pion, the ρ-meson or the K^0 are reproduced [MT00, AWW02, FW09]. One can also use u-, d- and s-quarks, fitting their masses to appropriate meson properties. In either case, one does not get the experimentally observed η and η' mesons in the ladder truncation. One will find the octet and singlet mesons η_8 and η_0 [AFW08]. There are two possibilities to include η-η'-splitting into the BSE-approach. One is to include terms connected to the axial anomaly using quark-gluon vertex dressings beyond the rainbow approximation (i.e. with a more complex structure than merely the tree-level) [AFW08]. The other is to include mixing of the flavour-singlet η_0-meson with the $J^{PC} = 0^{-+}$ pseudoscalar glueball, which should lead

[9] The glueball is a flavour singlet.

to an enhanced η_0-meson mass. Our set of BSEs (3.2.37) to (3.2.39) provides the simplest means of consistently including such a glueball/meson mixing into BSE calculations. Furthermore the scalar mesons, which are notoriously confusing, seemingly being a multiplet of several states even whose numbers are still under debate [N+10, vBRM+86] is subject to glueball mixing via the last diagram of (3.2.39). While in the pseudoscalar channel the mixing should not be too strong, for BSE calculations indicate a η_0 meson mass somewhere about 600 MeV to 700 MeV, while the mass of the pseudoscalar glueball is expected to be around 2500 MeV [CAD+06], in the scalar channel glueball/meson mixing is expected to be stronger. Comparing again the mass of a scalar meson in isospin limit obtained from BSE calculations, which is about 650 MeV, the mass of scalar $s\bar{s}$-mesons, which is found to be around 1000 MeV and the scalar glueball mass expected to be about 1700 MeV [CAD+06], sizable mixing is to be expected. Indeed the scalar mesons can hardly be described in terms of simple constituents, be it in non-relativistic quark models or a more sophisticated covariant and non-perturbative study using the BSE. Therefore the set of equations (3.2.37) to (3.2.39) may well provide a viable starting point for sophisticated investigations of a realistic meson spectrum using the combined Schwinger-Dyson/Bethe-Salpeter approach.

3.3 Representations of Bethe-Salpeter amplitudes

A convenient way to construct a representation for the Bethe-Salpeter amplitude is to find the most general scalar representation in the direct-product space of the constituent particles and then attach to it an angular momentum tensor that corresponds to the desired total spin J of the bound-state particle. The result can then be equipped with definite parity and charge parity (in case one deals with a particle-antiparticle system) properties by multiplying the appearing terms with some appropriate factors [LS69] as will be explicitly shown later.
In this section we first will recall a bit of formalism that is necessary for the construction of the basic scalar representations. We will do so for the rather general case that applies if the constituent particles are in the $(j,0)\oplus(0,j)$ representation. This kind of

representations can be used for a multitude of applications, most notably the BSEs describing mesons, which most conveniently employ the well-known Dirac spinors i.e. the $(1/2,0)\oplus(0,1/2)$ representation. After that we will study the $(1/2,1/2)$ representation which is the most appropriate for the treatment of glueballs and which we will use for our forthcoming investigations. Then the construction of the angular momentum tensor, which is to be multiplied with the scalar representation is reviewed. Finally we will recapitulate and assemble the formulae for arbitrary bound-state quantum numbers for constituent-particles in the $(1/2,0)\oplus(0,1/2)$ and the $(1/2,1/2)$ representation.

3.3.1 Basic invariant amplitudes

Spinor particles

In the following we will use the (A,B) representation formalism introduced in [Wei64a], which exhausts all finite dimensional representations of the Lorentz group.[10] For convenience of the reader we will restate the parts of the formalism, which are most important for the following. To specify any representation of the Lorentz group one starts with the infinitesimal Lorentz transformation

$$\Lambda^{\mu}{}_{\nu}=\delta^{\mu}{}_{\nu}+\omega^{\mu}{}_{\nu},$$

where $\omega_{\mu\nu}$ is antisymmetric. The unitary operator corresponding to this transformation is

$$U[1+\omega]=1+(i/2)\,J_{\mu\nu}\omega^{\mu\nu}.$$

One can group the operators $J_{\mu\nu}$ into two hermitian three vectors J_i and K_i being the generators of the rotations and the boosts respectively[11]

$$J_i=(1/2)\epsilon_{ijk}J_{jk},\quad K_i=J_{i0}.$$

[10] A pedagogical treatment of the (A,B) representations can be found in [Wei95].

[11] We work in four dimensions. Greek indices run from 0 to 3, while latin indices run from 1 to 3.

These operators satisfy the commutation relations

$$
\begin{aligned}
[J_i, J_j] &= i\epsilon_{ijk}\, J_k, \\
[J_i, K_j] &= i\epsilon_{ijk}\, K_k, \\
[K_i, K_j] &= -i\epsilon_{ijk}\, J_k,
\end{aligned}
$$

which can be grouped into a non-hermitian pair of generators

$$
\begin{aligned}
\mathbf{A} &= \frac{1}{2}\left(\mathbf{J} + i\mathbf{K}\right), \\
\mathbf{B} &= \frac{1}{2}\left(\mathbf{J} - i\mathbf{K}\right).
\end{aligned}
$$

The representation corresponding to these generators is called the (A, B) representation. It is $(2A+1)(2B+1)$-dimensional and irreducible. For any half-integer values of A and B one obtains the representation matrices by

$$
\begin{aligned}
\langle a, b|\, \mathbf{A}\, | a', b'\rangle &= \delta_{b\,b'}\, \mathbf{J}_{a\,a'}(A), \\
\langle a, b|\, \mathbf{B}\, | a', b'\rangle &= \delta_{a\,a'}\, \mathbf{J}_{b\,b'}(B),
\end{aligned}
$$

where a and b run from $-A$ to $+A$ and $-B$ to $+B$ respectively in unit steps. The matrices for $J(i)$ are the $2j+1$-dimensional representations of the rotation group

$$
\begin{aligned}
\left(J_x{}^{(j)} \pm i\, J_y{}^{(j)}\right)_{\sigma'\sigma} &= \delta_{\sigma',\sigma\pm1}\left[(j \mp \sigma)(j \pm \sigma + 1)\right]^{1/2}, \qquad (3.3.1)\\
\left(J_z{}^{(j)}\right)_{\sigma'\sigma} &= \delta_{\sigma'}\, \sigma,
\end{aligned}
$$

with σ and σ' running from $-j$ to $+j$ also in unit steps. The matrix representing a finite Lorentz transformation Λ is then obtained by exponentiating the generators with appropriate factors in the usual way and in the following will be denoted by $D[\Lambda]$. In this formalism a particle field of spin j can represented by a $(2j+1)$-component spinor in the $(j, 0)$ or $(0, j)$ representation. However if a field theory is to be parity-conserving it is necessary to include both representation types, which is usually done by combining them into a single field of type $(j, 0) \oplus (0, j)$. Obviously this representation is reducible and the Lorentz-transformations are

$$
\mathcal{D}^{(j)} = \begin{pmatrix} D^{(j,0)}[\Lambda] & 0 \\ 0 & D^{(0,j)}[\Lambda] \end{pmatrix}.
$$

By inspection one can see that $\mathcal{D}^{(j)}$ for $j = 1/2$ reduces to the well-known Dirac spinor formalism. One should note that this formalism deals with massive fields, but it can also be applied to massless fields with a few modifications [Wei64b].
To construct the scalar Bethe-Salpeter amplitude for two particles in the $(j,0) \oplus (0,j)$ representation, once again recall that such an amplitude can be be constructed from the direct product of the representations of the two constituent particles. Therefore its matrix representation has to be $(2(2j+1)) \times (2(2j+1))$-dimensional. Since we look for the most general covariant decomposition of the Bethe-Salpeter amplitude, we want a complete set of matrices satisfying

$$\mathcal{D}^{(j)}[\Lambda]\,\gamma^{\mu_1\mu_2\ldots\mu_{2j}}\,\left(\mathcal{D}^{(j)}[\Lambda]\right)^{-1} = \Lambda_{\nu_1}{}^{\mu_1}\,\Lambda_{\nu_2}{}^{\mu_2}\ldots\Lambda_{\nu_{2j}}{}^{\mu_{2j}}\,\gamma^{\nu_1\nu_2\ldots\nu_{2j}}. \tag{3.3.2}$$

It is clear that such a set is nothing but a generalised version of the Dirac gamma-matrices. It was first introduced in [Wei64a], where also the explicit construction of these matrices can be found.
Being somewhat lengthy we will not repeat the construction here. Rather we will discuss some important properties of those matrices. From (3.3.2) it is clear that the generalised gamma-matrices form a Lorentz-tensor $\gamma^{\mu_1\mu_2\ldots\mu_{2j}}$. It is symmetric in its Lorentz indices and traceless with respect to any pair of Lorentz indices[12]

$$\gamma^{\mu_1\mu_2\ldots\mu_{2j}} = \gamma^{\mathcal{P}[\mu_1\mu_2\ldots\mu_{2j}]},$$
$$g_{\mu_k\mu_l}\gamma^{\mu_1\mu_2\ldots\mu_k\ldots\mu_l\ldots\mu_{2j}} = 0.$$

In four dimension any symmetric and traceless tensor of rank k has $(k+1)^2$ independent components. Since we are looking for a complete set of matrices, in the $(j,0) \oplus (0,j)$ representation, which is $2(2j+1)$-dimensional, we need $(2(2j+1))^2$ independent matrices and from the basic set $\gamma^{\mu_1\mu_2\ldots\mu_{2j}}$ we have $(2j+1)^2$. The missing matrices can be found by constructing higher rank tensors out of the $\gamma^{\mu_1\mu_2\ldots\mu_{2j}}$ by antisymmetrised products

$$\sigma_{(k)}^{\mu_1\mu_2\ldots\mu_{2k-1}\mu_{2k}} = \sum_{\mathcal{P}} sign[\mathcal{P}]\,\mathcal{P}[\gamma^{\mu_1\mu_2}\ldots\gamma^{\mu_{2k-1}\mu_{2k}}]. \tag{3.3.3}$$

However this method can turn out to be quite tedious. In order to reduce the number of matrices that have to be calculated by antisymmetric products we note, that in such

[12] $\mathcal{P}[\ldots]$ denotes any permutation of the indices in brackets.

a complete set of matrices there is a *pseudoscalar* matrix included that is conventionally denoted as γ_5, which can be chosen as

$$\gamma_5 = \begin{pmatrix} Id_{2j+1} & 0 \\ 0 & -Id_{2j+1} \end{pmatrix}$$

and a *scalar* matrix which is nothing but the identity matrix. To get another part of the complete set of matrices, we take the product $\gamma_5\gamma^{\mu_1\mu_2\cdots\mu_{2j}}$. With the scalar and pseudoscalar matrix we then have $2(1+(2j+1)^2)$ independent matrices. The others have to be calculated with antisymmetric products (3.3.3). With a complete set of covariant matrices at hand, one can now construct invariants by contracting the matrix-valued tensors and the momentum four-vectors that appear on the LHS of the BSE. It is usually convenient to choose the rest-frame and the two momenta to be the center-of-mass momentum t_μ and the relative momentum r_μ between one constituent particle and center-of-mass momentum. For the case $j = 1/2$ we get the invariants

$$\begin{array}{lllll} Id_4, & t_\mu\gamma^\mu, & r_\mu\gamma^\mu, & t_\mu r_\nu\sigma^{\mu\nu}, & (scalar) \\ \gamma_5, & t_\mu\gamma_5\gamma^\mu, & r_\mu\gamma_5\gamma^\mu, & t_\mu r_\nu\gamma_5\sigma^{\mu\nu}. & (pseudoscalar) \end{array} \tag{3.3.4}$$

The most general invariant is therefore a linear combination of these possible invariants. The factors can in general be functions of the momenta appearing in the BSE. Usually one is interested in amplitudes which do not only transform according to a spin 0 representation, but additional have a definite parity. These are obtained by restricting the set of invariants (3.3.4) to the scalars or pseudoscalars, since the angle between the total and the relative momentum does not change under parity. So the most general scalar amplitude is

$$\Gamma_{0^+}(t^2,r^2,\theta) = A(t^2,r^2,\theta)\,Id_4 + B(t^2,r^2,\theta)\not{r} + C(t^2,r^2,\theta)\not{t} + D(t^2,r^2,\theta)\,t_\mu r_\nu\sigma^{\mu\nu}, \tag{3.3.5}$$

where A, B, C, D are scalar functions and θ is the angle between the momenta r and t. Obviously the pseudoscalar amplitude is obtained by multiplying (3.3.5) with γ_5.

If one is interested in the properties of particle-antiparticle bound states also the charge-conjugation parity has to be fixed. In the case of a $(j,0)\oplus(0,j)$ spinor formalism, there is a natural charge-conjugation matrix $\mathcal{C}$ included in the set of generalised gamma-matrices defined by

$$\mathcal{C} = \begin{pmatrix} C & 0 \\ 0 & C \end{pmatrix}, \tag{3.3.6}$$

with C satisfying

$$\begin{aligned} C\mathbf{J}^{(j)}C^{-1} &= -\mathbf{J}^{(j)*}, \\ C^*C = (-1)^{2j}\, Id_{2j+1}, &\quad C^\dagger C = Id_{2j+1}, \end{aligned} \tag{3.3.7}$$

where $\mathbf{J}^{(j)}$ is the vector of spin-matrices (3.3.1). To construct an amplitude that is an eigenstate (modulo matrix transposition) under charge-conjugation, one firstly has to inspect the behaviour of the invariants under charge-conjugation. For instance invariants of the $j = 1/2$ scalar (3.3.5) transform like[13]

$$\begin{aligned} \mathcal{C} Id_4 \mathcal{C}^{-1} &\to Id_4, \\ \mathcal{C}\not{r}\mathcal{C}^{-1} &\to \not{r}^T, \\ \mathcal{C}\not{t}\mathcal{C}^{-1} &\to -\not{t}^T, \\ \mathcal{C} t_\mu r_\nu \sigma^{\mu\nu} \mathcal{C}^{-1} &\to \left(t_\mu r_\nu \sigma^{\mu\nu}\right)^T, \end{aligned}$$

since $\mathcal{C}\gamma^\mu\mathcal{C}^{-1} \to (-\gamma^\mu)^T$ and also the relative momentum changes because charge-conjugation exchanges particles and antiparticles thus reverses the spin lines $r_\mu \to -r_\mu$. The latter however also gives a constraint to the coefficient-functions A, B, C, D. Since by charge-conjugation the scalar-products $(l \cdot t)$ between the loop-momentum and the total momentum l on the RHS of the BSE also pick up a minus, the coefficient-functions have to be either even or odd in $(l \cdot t)$. In turn this yields a constraint on the possible approximations to the scattering kernel K in (3.2.1). A kernel that is suitable for the description of particle-antiparticle bound states should have a definite behaviour under charge-conjugation. In the simplest case the kernel itself is even in $(l \cdot t)$ and the charge-conjugation is determined by the structure of the amplitude only. Then the behaviour under charge-conjugation can be fixed by multiplying the invariants with appropriate

[13] All transpositions refer to the Dirac indices.

factors of $(r \cdot t)$. For instance the scalar (3.3.5) with even charge-conjugation parity reads

$$\Gamma_{0^{++}}(t^2, r^2, \theta) = A(t^2, r^2, \theta)\, Id_4 + B(t^2, r^2, \theta) \not{r} \\ + C(t^2, r^2, \theta)\ (r \cdot t) \not{t} + D(t^2, r^2, \theta)\, t_\mu r_\nu \sigma^{\mu\nu} \tag{3.3.8}$$

and transforms under charge-conjugation like $\mathcal{C}\Gamma^{0^{++}}(t^2, r^2, \theta)\mathcal{C}^{-1} \rightarrow (\Gamma_{0^{++}}(t^2, r^2, \theta))^T$. Keeping in mind that for any two square matrices $A^T B^T = (B\, A)^T$ one finds for the pseudoscalar invariants

$$\begin{aligned} \mathcal{C}\gamma_5\mathcal{C}^{-1} &\rightarrow (\gamma_5)^T\,, \\ \mathcal{C}\gamma_5 \not{r} \mathcal{C}^{-1} &\rightarrow -(\gamma_5 \not{r})^T\,, \\ \mathcal{C}\not{t}\mathcal{C}^{-1} &\rightarrow (\gamma_5 \not{t})^T\,, \\ \mathcal{C} t_\mu r_\nu \sigma^{\mu\nu} \mathcal{C}^{-1} &\rightarrow (\gamma_5\, t_\mu r_\nu \sigma^{\mu\nu})^T\,. \end{aligned}$$

Thus the pseudoscalar amplitude reads

$$\Gamma^{0^{-+}}(t^2, r^2, \theta) = \gamma_5 \Bigg(A(t^2, r^2, \theta)\, Id_4 + B(t^2, r^2, \theta)\ (r \cdot t) \not{r} \\ + C(t^2, r^2, \theta) \not{t} + D(t^2, r^2, \theta)\, t_\mu r_\nu \sigma^{\mu\nu} \Bigg). \tag{3.3.9}$$

These two amplitudes are the basis for the construction of the other states. The behaviour under charge-conjugation can be flipped by simply multiplying with a factor of $(r \cdot t)$. For representations with higher j the procedure is analogous. There will be more possible invariants since the matrix-valued tensors $\gamma^{\mu_1\mu_2\ldots\mu_{2j}}$ have more indices. However because there are only two momentum four-vectors present, not all matrix-valued tensors constructed from antisymmetrised products (3.3.3) will contribute.

Particles in the $(1/2, 1/2)$ representation

The above procedure for some of the possible states simplifies significantly in the $(1/2, 1/2)$ representation. It is the representation which will be used in the following to describe Glueballs via the BSE formalism. In this case one basic invariant is readily found. Since the constituent particles are described by the well-known traceless, self-adjoint and antisymmetric field-strength tensors $F^{\mu\nu}$, the corresponding propagators are also rank two Lorentz-tensors $D^{\mu\nu}$. It is therefore obvious that the basic scalar Bethe-Salpeter amplitude in this case is a three-point function with two Lorentz-indices. To find the basic scalar invariant, we look for all rank two Lorentz-tensors, that are invariant under Lorentz-transformations

$$\Lambda^{\kappa}{}_{\mu}\,\Lambda^{\lambda}{}_{\nu}\,T_{\kappa\lambda} = T_{\mu\nu}.$$

However there are only two such tensors. One is the metric tensor $g_{\mu\nu}$ and the other can be build from a Levi-Civita tensor contracted with two different momenta $p^{\kappa}\,q^{\lambda}\,\epsilon_{\kappa\lambda\mu\nu}$. Choosing the momenta p and q to be the momenta appearing in the Bethe-Salpeter amplitude r and t, these two tensors give the basic scalar and pseudoscalar invariants respectively. So we have

$$\Gamma^{0++}_{\mu\nu}(t^2, r^2, \theta) = A(t^2, r^2, \theta)\, g_{\mu\nu} \tag{3.3.10}$$

and

$$\Gamma^{0-+}_{\mu\nu}(t^2, r^2, \theta) = A(t^2, r^2, \theta)\ (r \cdot t)\ r^{\kappa} t^{\lambda}\,\epsilon_{\kappa\lambda\mu\nu}. \tag{3.3.11}$$

However these are not yet all possible basic tensors. It is known that the combination of two massless vector particles give a scalar basic amplitude and a spin two tensor besides[14]

[14] At this place a remark about the comparability of the representations for Bethe-Salpeter amplitudes, we construct here and the amplitudes one can construct from spinors and vectors etc. which are much more commonly found in the literature, seems in order.
It is a well-known text-book excercise in quantum mechanics to construct amplitudes from direct products of e.g. spinors that can be used in calculations for instance of atoms with several electrons. The most commonly known example is possibly the singlet-triplet construction of two Fermi-spinors. It is as well possible to perform that construction using Dirac-spinors. The result obviously is a 4×4 matrix in Dirac-space, which transforms like a bispinor. This is the same kind of object as we have constructed above (cf. equation (3.3.2)). However one can check explicitly that both representations are not equivalent i.e. there is no linear map between them. That futhermore means that it is not possible to multiply some momentum vectors into the bispinor, constructed as a direct product from two Dirac-spinors, to find our set of basic invariants (like we have in equation (3.3.4)) and then

[Lan48]. Indeed there is another basic invariant amplitude that can be constructed which has spin two. It can be written explictly (and somewhat lengthy) as

$$\begin{aligned}\Gamma^2_{\mu\nu\mu_1\mu_2} =& P^4\Big(-\frac{1}{3}g_{\mu\nu}g_{\mu_1\mu_2} + \frac{1}{2}g_{\mu\mu_1}g_{\nu\mu_2} + \frac{1}{2}g_{\mu\mu_2}g_{\nu\mu_1}\Big) \\ &+ P^2\Big(\frac{1}{3}g_{\mu\nu}P_{\mu_1}P_{\mu_2} - \frac{1}{2}g_{\mu\mu_1}P_\nu P_{\mu_2} - \frac{1}{2}g_{\mu\mu_2}P_\nu P_{\mu_1} \\ &\quad - \frac{1}{2}g_{\nu\mu_1}P_\mu P_{\mu_2} - \frac{1}{2}g_{\nu\mu_2}P_{\mu\mu_1} + \frac{1}{3}g_{\mu_1\mu_2}P_\mu P_\nu\Big) \\ &+ \frac{2}{3}P_\mu P_\nu P_{\mu_1}P_{\mu_2}.\end{aligned} \tag{3.3.12}$$

For states with spin two or more accordingly there can be two amplitudes contributing. However due to the mixing of the basic indices μ and ν (to which the constituent particles connect) and the spin indices μ_1 and μ_2, the distinction between basic and angular part of the amplitude is obscured. The structure (3.3.12) obeys the spin two conditions in every pair of indices. The construction of states with total spin higher than $J = 2$ will become quite complicated yet possible because of this structure of (3.3.12). We will come back to that in the next chapter.

In the $(1/2, 1/2)$ representation there is no matrix that acts as a charge-conjugation operator. This is reminiscent of the fact that the particles in this representation are their own antiparticles. Therefore the only thing a charge-conjugation operation does to the BSE is to reverse the routing of the loop momenta on the RHS. To account for this, in (3.3.11) a factor of $(r \cdot t)$ has been introduced.

---

attach the angular momentum tensors we will construct in the next section, to find a representation of an amplitude with arbitrary quantum numbers. The same is true in other representations of the constituents. For instance we will not obtain the basic invariants in the $(1/2, 1/2)$ representation by constructing direct products of Lorentz-vectors.

However looking at the direct-products can give valuable hints on the existence of further states. Note also that the direct-product construction being non- or semi-relativistic can give further restrictions on possible amplitudes, which are not obvious or even absent in our explicitely covariant approach. In the constituent quark model of mesons for instance there are significantly less bound states that can in principle be formed than in our approach e.g. the 0^{+-} or the 1^{-+}, which are denoted as being *exotic* from the constituent quark model perspective.

The same is true for the $(1/2, 1/2)$ representation. We can easily form for instance vector states, which are forbidden [Lan48, Yan50]. We will use the information coming from the direct-product construction as selection rules in section 5.1.

3.3.2 Angular momentum tensors

Having constructed the basic invariant amplitudes in the previous section, the full amplitude for higher total angular momentum can be constructed by multiplying one of the basic invariant amplitudes with a tensor representing a given total angular momentum J. Such a tensor is required to have precisely $2J+1$ independent components to represent the possible spin polarisations. Note that number of independent components of a Lorentz-tensor is invariant under Lorentz-transformations. The construction of such tensors is known and a detailed treatment can be found e.g. in [Zem65]. We will repeat parts of the construction here in a slightly more explicit form focused directly on the construction of Bethe-Salpeter amplitudes. Consider first tensors $T_{a_1,\ldots,a_J}$ in three-space of rank J. To represent angular momentum J we require the tensor to be

1. symmetric in all indices
$$T_{a_1\ldots a_J} = T_{\mathcal{P}[a_1\ldots a_J]} \tag{3.3.13}$$

2. traceless with respect to any pair of indices
$$\sum_m T_{\ldots m \ldots m \ldots} = 0. \tag{3.3.14}$$

Constraint (3.3.13) leaves the tensor with $\frac{1}{2}\left(J^2+3J+2\right)$ independent components, while (3.3.14) imposes $\frac{1}{2}\left(J^2-J\right)$ further restrictions, thus leading to a tensor with $2J+1$ independent components as required. It is also possible to construct tensors representing spin $j+\frac{1}{2}$ with integer j. Since we are concerned with Bethe-Salpeter equations of two particles in the same representation, so that the total angular momentum is integer, we will not pursue that possibility but instead refer the reader again to [Zem65]. The construction of tensors in three-space is now easily transferred to four-tensors. If we require the tensor $T_{\mu_1\ldots\mu_J}$ to be transverse to the total momentum of the particle in every index $t^\nu T_{\ldots\nu\ldots} = 0$ and adopt the particles rest-frame, we see that all components with time-like indices vanish, leaving only components with space-like indices. So we are left with nothing but the three-tensor considered before, which has $2J+1$ independent components as desired. Thus we find the constraints for a Lorentz-tensor $T^J_{\mu_1\ldots\mu_J}$ of rank J to represent angular momentum J:

1. T is symmetric in all indices

$$T^J_{\mu_1\cdots\mu_J} = T^J_{\mathcal{P}[\mu_1\cdots\mu_J]} \tag{3.3.15}$$

2. T is transverse to the total momentum of the particle in every index

$$t^\nu \, T^J_{\dots\nu\dots} = 0 \tag{3.3.16}$$

3. T is traceless in every pair of indices in the rest-frame

$$T^{J,\dots\lambda\dots}_{\dots\lambda\dots} = 0. \tag{3.3.17}$$

The idea to construct such a tensors of angular momentum J is to build the J-fold tensor product of a transverse projector that transforms like a vector and then subtract the traces with respect to every pair of indices. Consider $J = 1$. A suitable transverse projector[15] can be obtained by contracting the standard tensor transverse projector with a momentum[16]

$$Q_\mu = \left(g_{\mu\nu} - \frac{t_\mu \, t_\nu}{t^2} \right) r^\nu = \left(r_\mu - \frac{(r \cdot t) \; t_\mu}{t^2} \right),$$

where we already chose the momentum to be the relative momentum appearing in the Bethe-Salpeter amplitude. However, if we work in a $(j,0) \oplus (0,j)$ representation, this is not the only possible tensor. It is also possible to contract the standard transverse projector with an appropriate vector-like spinorial object. In the $j = 1/2$ this object is just the vector of gamma matrices

$$T^{(j=1/2)}_\mu = \left(g_{\mu\nu} - \frac{t_\mu \, t_\nu}{t^2} \right) \gamma^\nu = \left(\gamma_\mu - \frac{\not{t} \; t_\mu}{t^2} \right).$$

For higher j the tensor of gamma matrices $\gamma_{\mu_1\dots\mu_{2j}}$ has to be contracted with $2j - 1$ momentum vectors. In principle these momenta can be chosen arbitrarily. For technical reasons it is however convenient to choose all momenta to be the total momentum, since otherwise on the RHS of the BSE further powers of the loop-momentum are introduced,

[15] More precisely these are pseudoprojectors, since they are not normalised

[16] The relative momentum r is replaced by l, in an amplitude appearing on the RHS of the BSE.

which have to be integrated over in principle up to infinity. These can lead to accuracy problems, when the BSE is solved numerically (which usually is the case). Furthermore the tensor-algebra which is introduced in the following is simplified a lot, when there is only one momentum present in the calculations. However choosing a different momentum does not change anything fundamental. So in the $j = 1$ case a convenient choice for the projector is

$$T_\mu^{(j=1)} = \left(g_{\mu\nu} - \frac{t_\mu \, t_\nu}{t^2}\right) \gamma^{\nu\rho} \, t_\rho.$$

And analogously the vector transverse projectors can be constructed for higher j. Since having only one Lorentz-index these vector transverse projectors are trivially traceless and also symmetric. So they already are the possible angular momentum tensors for $J = 1$. For higher J one builds symmetrised J-fold tensor products of the projectors[8]

$$\tilde{Q}_{\mu_1 \dots \mu_J} = \frac{1}{J!} \, Q_{\{\mu_1} \dots Q_{\mu_J\}}, \tag{3.3.18}$$

$${}^{(n)}\tilde{T}^{(j)}_{\mu_1 \dots \mu_J} = \frac{1}{(J-n-1)!} \, T^{(j)}_{\{\mu_1} \dots T^{(j)}_{\mu_n} \, Q_{\mu_{n+1}} \dots Q_{\mu_J\}}. \tag{3.3.19}$$

The next step is to remove the traces of these tensors with respect to every pair of indices. However this is a very tedious thing to do, if one deals with $(j,0) \oplus (0,j)$ representations with $j > 1/2$ and angular momentum tensors with $J > 1$, since the gamma matrix algebra gets far more complicated than the rather simple Clifford algebra of the $j = 1/2$ case. Another complication arises from the fact that for $j = 1/2$, most of the tensors (3.3.19) vanish, when the traces are removed, which is not the case for $j > 1/2$. To see this let us concentrate on the $j = 1/2$ case and note that[17]

$$\{T_\mu, T_\nu\} = 2\tau_{\mu\nu},$$

where $\tau_{\mu\nu} = \left(g_{\mu\nu} - \frac{t_\mu t_\nu}{t^2}\right)$ is the standard transverse projector with respect to the total momentum t. However the standard transverse projector is manifestly not traceless in the rest-frame and neither is any tensor-product build from it. Since the tensor-product (3.3.19) for every $n > 1$ can be expanded in terms of anticommutators $\{T_\mu, T_\nu\}$ it is clear

[8] The curly brackets around the indices indicated total symmetrisation with respect to these indices.

[17] For convenience we omit the representation index from now on and write $T_{\mu_1 \dots \mu_J}$ instead of $T^{(1/2)}_{\mu_1 \dots \mu_J}$.

that these all vanish, when tracelessness is required. Thus only tensors containing exactly one T_μ do not vanish. We thus find that all angular momentum tensors that will appear are

$$\tilde{Q}_{\mu_1\cdots\mu_J} = Q_{\mu_1}\ldots Q_{\mu_J}, \tag{3.3.20}$$

$$\tilde{T}_{\mu_1\cdots\mu_J} = \frac{1}{(J-1)!}\, T_{\{\mu_1}\, Q_{\mu_2}\ldots Q_{\mu_J\}}. \tag{3.3.21}$$

Now we have to make the tensors (3.3.20) and (3.3.21) traceless. The method to do this can be found in [Zem65] and is applied here explicitly. The general formula for a symmetric and traceless tensor build from a totally symmetric raw tensor $\tilde{T}$ is

$$\begin{aligned} T_{\mu_1\cdots\mu_J} =& \tilde{T}_{\mu_1\cdots\mu_J} - (2J-1)^{-1} \sum_{P_{\mu_k}} \delta_{\mu_1\mu_2}\, \tilde{T}^{\kappa}{}_{\kappa\mu_3\cdots\mu_J} + \\ & (2J-1)^{-1}\,(2J-3)^{-1} \sum_{P_{\mu_k}} \delta_{\mu_1\mu_2}\,\delta_{\mu_3\mu_4}\, \tilde{T}^{\kappa\lambda}{}_{\kappa\lambda\mu_5\cdots\mu_J} - \ldots etc, \end{aligned} \tag{3.3.22}$$

where $\sum_P$ denotes the sum over all essentially different permutations of the indices.[18] Note that $\delta_{\mu\nu}$ is the space-like Kronecker-delta in the rest-frame of the particle. Written in Lorentz-covariant form it is just the transverse projector to the total momentum t

$$\delta_{\mu\nu} = \tau_{\mu\nu}.$$

Furthermore let us define

$$f_1 = \not{r} - \frac{(r\cdot t)\not{t}}{t^2}, \tag{3.3.23}$$

$$f_2 = r^2 - \frac{(r\cdot t)^2}{t^2} \tag{3.3.24}$$

and the tensors[19]

$$\begin{aligned} A^{J,j}_{\mu_1\cdots\mu_J} &= 2j\, f_1\, f_2^{(j-1)}\, \delta_{\{\mu_1\mu_2}\ldots\delta_{\mu_{2j-1}\,\mu_{2j}}\, \tilde{Q}^{(J-2j)}_{\mu_{2j+1}\cdots\mu_J\}} \\ &\quad + f_2^j\, \delta_{\{\mu_1\mu_2}\ldots\delta_{\mu_{2j-1}\,\mu_{2j}}\, \tilde{T}^{(J-2j)}_{\mu_{2j+1}\cdots\mu_J\}}, \qquad J > 2j \end{aligned} \tag{3.3.25}$$

$$A^{J,j}_{\mu_1\cdots\mu_J} = J\, f_1\, f_2^{(J/2-1)}\, \delta_{\{\mu_1\mu_2}\ldots\delta_{\mu_{J-1}\,\mu_J\}} \qquad J = 2j \tag{3.3.26}$$

$$B^{J,j}_{\mu_1\cdots\mu_J} = f_2^j\, \delta_{\{\mu_1\mu_2}\ldots\delta_{\mu_{2j-1}\,\mu_{2j}}\, \tilde{Q}^{(J-2j)}_{\mu_{2j+1}\cdots\mu_J\}}, \qquad J > 2j \tag{3.3.27}$$

$$B^{J,j}_{\mu_1\cdots\mu_J} = f_2^{J/2}\, \delta_{\{\mu_1\mu_2}\ldots\delta_{\mu_{J-1}\,\mu_J\}}. \qquad J = 2j \tag{3.3.28}$$

[18] That means that the sum has to be divided by appropriate combinatorical factors.

[19] For the convenience of the reader we have denoted the rank of the raw tensors as a superscript.

Using the raw tensors (3.3.18) and (3.3.21) the final formula for the angular momentum tensors of arbitrary integer J in the $(1/2,0)\oplus(0,1/2)$ representation are

$$T_{\mu_1\ldots\mu_J}=\tilde{T}_{\mu_1\ldots\mu_J}+\sum_{j=1}^{J\geq 2j}(-1)^j\,\frac{1}{j!\,2^j}\left(\prod_{k=1}^{j}2(J-k)+1\right)^{-1}A^{J,j}_{\mu_1\ldots\mu_J}, \tag{3.3.29}$$

$$Q_{\mu_1\ldots\mu_J}=\tilde{Q}_{\mu_1\ldots\mu_J}+\sum_{j=1}^{J\geq 2j}(-1)^j\,\frac{1}{j!\,2^j}\left(\prod_{k=1}^{j}2(J-k)+1\right)^{-1}B^{J,j}_{\mu_1\ldots\mu_J}. \tag{3.3.30}$$

For the $(1/2,1/2)$ representation there are no gamma matrices, thus (3.3.30) is the only possible angular momentum tensor, furthermore simplifying matters a lot.

3.3.3 Bethe-Salpeter amplitudes for any J

The final step is to combine the angular momentum tensors and basic scalar or pseudoscalar amplitudes. If no definite behaviour under charge-conjugation is required this is trivial. We multiply the angular momentum tensor (3.3.29) and (3.3.30) with appropriate J and the basic scalar (3.3.5).
For the pseudoscalar amplitude we additionally multiply with γ_5

$$\begin{aligned}
&\Gamma^{J+}_{\mu_1\ldots\mu_J}(t^2,r^2,\theta)=\\
&\quad T_{\mu_1\ldots\mu_J}\left(A(t^2,r^2,\theta)\,Id_4+B(t^2,r^2,\theta)\not{r}+C(t^2,r^2,\theta)\not{t}+D(t^2,r^2,\theta)\,t_\mu r_\nu\sigma^{\mu\nu}\right)\\
&\quad+Q_{\mu_1\ldots\mu_J}\left(E(t^2,r^2,\theta)\,Id_4+F(t^2,r^2,\theta)\not{r}+G(t^2,r^2,\theta)\not{t}+H(t^2,r^2,\theta)\,t_\mu r_\nu\sigma^{\mu\nu}\right),
\end{aligned} \tag{3.3.31}$$

$$\begin{aligned}
&\Gamma^{J-}_{\mu_1\ldots\mu_J}(t^2,r^2,\theta)=\\
&\quad\gamma_5\,T_{\mu_1\ldots\mu_J}\left(A(t^2,r^2,\theta)\,Id_4+B(t^2,r^2,\theta)\not{r}+C(t^2,r^2,\theta)\not{t}+D(t^2,r^2,\theta)\,t_\mu r_\nu\sigma^{\mu\nu}\right)\\
&\quad+Q_{\mu_1\ldots\mu_J}\,\gamma_5\left(E(t^2,r^2,\theta)\,Id_4+F(t^2,r^2,\theta)\not{r}+G(t^2,r^2,\theta)\not{t}+H(t^2,r^2,\theta)\,t_\mu r_\nu\sigma^{\mu\nu}\right).
\end{aligned} \tag{3.3.32}$$

To find the amplitudes with definite charge-conjugation parity the following identities for the behaviour of the angular momentum tensor under charge-conjugation are useful

$$\mathcal{C}T_{\mu_1\cdots\mu_J}\mathcal{C}^{-1} = (-1)^J \left(T_{\mu_1\cdots\mu_J}\right)^T, \tag{3.3.33}$$

$$\mathcal{C}Q_{\mu_1\cdots\mu_J}\mathcal{C}^{-1} = (-1)^J \, Q_{\mu_1\cdots\mu_J}, \tag{3.3.34}$$

$$\mathcal{C}\left\{T_{\mu_1\cdots\mu_J}, \not r\right\}\mathcal{C}^{-1} = 2JQ_{\mu_1\cdots\mu_J}, \tag{3.3.35}$$

$$\mathcal{C}\left\{T_{\mu_1\cdots\mu_J}, \not t\right\}\mathcal{C}^{-1} = 0, \tag{3.3.36}$$

$$\mathcal{C}\left[T_{\mu_1\cdots\mu_J}\left[\not r, \not t\right]\right]\mathcal{C}^{-1} = 4JQ_{\mu_1\cdots\mu_J}\not t^T, \tag{3.3.37}$$

$$\begin{aligned} \mathcal{C}\left[Q_{\mu_1\cdots\mu_J}, \not r\right]\mathcal{C}^{-1} &= \mathcal{C}^{-1}\left[Q_{\mu_1\cdots\mu_J}, \not t\right]\mathcal{C}^{-1}, \\ &= \mathcal{C}^{-1}\left[Q_{\mu_1\cdots\mu_J}, \left[\not r, \not t\right]\right]\mathcal{C}^{-1} \\ &= 0. \end{aligned} \tag{3.3.38}$$

We multiply the angular momentum tensors with the scalar (3.3.8) and pseudoscalar (3.3.9) amplitudes respectively. By inspection it is found that the resulting amplitudes generate an additional term under charge-conjugation. After an appropriate subtraction for even J we find[20]

$$\begin{aligned} \Gamma^{J++}_{\mu_1\cdots\mu_J} &= \\ & T_{\mu_1\cdots\mu_J}\left(A\, Id_4 + B\,(r\cdot t)\not r + C\not t + D\, t_\mu r_\nu \sigma^{\mu\nu}\right) \\ & \qquad - J\, Q_{\mu_1\cdots\mu_J}\left((r\cdot t)\, B + 2\, D\not t\right) \\ & \quad + Q_{\mu_1\cdots\mu_J}\left(E\, Id_4 + F\not r + G\,(r\cdot t)\not t + H\, t_\mu r_\nu \sigma^{\mu\nu}\right), \end{aligned} \tag{3.3.39}$$

$$\begin{aligned} \Gamma^{J--}_{\mu_1\cdots\mu_J} &= \\ & \gamma_5\, T_{\mu_1\cdots\mu_J}\left(A\, Id_4 + B\not r + C\,(r\cdot t)\not t + D\, t_\mu r_\nu \sigma^{\mu\nu}\right) \\ & \qquad - J\,\gamma_5\, Q_{\mu_1\cdots\mu_J}\left(B + 2\, D\not t\right) \\ & \quad + Q_{\mu_1\cdots\mu_J}\,(r\cdot t)\,\gamma_5\left(E\, Id_4 + F\,(r\cdot t)\not r + G\not t + H\, t_\mu r_\nu \sigma^{\mu\nu}\right) \end{aligned} \tag{3.3.40}$$

[20] We omit the function arguments for brevity.

and similarly for odd J

$$
\begin{aligned}
\Gamma^{J--}_{\mu_1\dots\mu_J} = \\
& T_{\mu_1\dots\mu_J}\left(A\, Id_4 + B\,(r\cdot t)\not{r} + C\not{t} + D\, t_\mu r_\nu \sigma^{\mu\nu}\right) \\
& \quad - J\, Q_{\mu_1\dots\mu_J}\left((r\cdot t)\; B + 2\, D\not{t}\right) \\
& + Q_{\mu_1\dots\mu_J}\left(E\, Id_4 + F\not{r} + G\,(r\cdot t)\not{t} + H\, t_\mu r_\nu \sigma^{\mu\nu}\right), \qquad (3.3.41) \\
\Gamma^{J++}_{\mu_1\dots\mu_J} = \\
& \gamma_5\, T_{\mu_1\dots\mu_J}\left(A\, Id_4 + B\not{r} + C\,(r\cdot t)\not{t} + D\, t_\mu r_\nu \sigma^{\mu\nu}\right) \\
& \quad - J\,\gamma_5\, Q_{\mu_1\dots\mu_J}\left(B + 2\, D\not{t}\right) \\
& + Q_{\mu_1\dots\mu_J}\,(r\cdot t)\;\gamma_5\left(E\, Id_4 + F\,(r\cdot t)\not{r} + G\not{t} + H\, t_\mu r_\nu \sigma^{\mu\nu}\right). \qquad (3.3.42)
\end{aligned}
$$

Obviously for even and odd J the parity and charge-parity properties of the amplitudes just interchange, which is due to (3.3.33) and (3.3.34). As before the behaviour under charge-conjugation can be flipped by multiplying the amplitudes with an overall factor of $(r\cdot t)$.

As already mentioned in the $(1/2,1/2)$ representation the only possible angular momentum tensor is $Q_{\mu_1\dots\mu_J}$. Multiplying this tensor with the amplitudes (3.3.10) and (3.3.11) respectively yields Bethe-Salpeter amplitudes for arbitrary even J

$$\Gamma^{J++}_{\mu\nu,\mu_1\dots\mu_J}(t^2,r^2,\theta) = Q_{\mu_1\dots\mu_J}\, A(t^2,r^2,\theta)\, g_{\mu\nu}, \qquad (3.3.43)$$

$$\Gamma^{J-+}_{\mu\nu,\mu_1\dots\mu_J}(t^2,r^2,\theta) = Q_{\mu_1\dots\mu_J}\, A(t^2,r^2,\theta)\;(r\cdot t)\; r^\kappa t^\lambda\, \epsilon_{\kappa\lambda\mu\nu} \qquad (3.3.44)$$

and if J is odd

$$\Gamma^{J++}_{\mu\nu,\mu_1\dots\mu_J}(t^2,r^2,\theta) = Q_{\mu_1\dots\mu_J}\, A(t^2,r^2,\theta)\;(r\cdot t)\; r^\kappa t^\lambda\, \epsilon_{\kappa\lambda\mu\nu}. \qquad (3.3.45)$$

$$\Gamma^{J-+}_{\mu\nu,\mu_1\dots\mu_J}(t^2,r^2,\theta) = Q_{\mu_1\dots\mu_J}\, A(t^2,r^2,\theta)\, g_{\mu\nu}. \qquad (3.3.46)$$

However as mentioned in section 3.3.1 there is another basic invariant $\Gamma^2_{\mu\nu\mu_1\mu_2}$ (cf.. (3.3.12)), which has spin two. This structure contributes to states higher than $J=1$. In case of spin two it can only contribute to the states 2^{++} and 2^{+-}, for its parity cannot be flipped, because it already has four indices and the parity flipping pseudoscalar tensor

(3.3.11) would add indices. For higher states than $J = 2$ the construction becomes awkward, when $\Gamma^2_{\mu\nu\mu_1\mu_2}$ contributes. The respective raw tensor from which the represention of a spin J state is constructed, is in principle also obtained by multiplying $\Gamma^2_{\mu\nu\mu_1\mu_2}$ with $J-2$ tensors Q_μ and is then totally symmetrised. The result obvioulsy is symmetric and transverse to the total momentum in all indices. It then is made traceless according to (3.3.22). However due to the total symmetry of all indices of $\Gamma^2_{\mu\nu\mu_1\mu_2}$ the result becomes lengthy and it is hard to give a closed formula for any J. The tensor $\Gamma^2_{\mu\nu\mu_1\mu_2}$ does not contribute to states with spin zero or one. Obviously the simplest example of a state containing a contribution of $\Gamma^2_{\mu\nu\mu_1\mu_2}$ is

$$\Gamma^{2++}_{\mu\nu,\mu_1\mu_2}(t^2,r^2,\theta) = Q_{\mu_1\mu_2}\, A(t^2,r^2,\theta)\, g_{\mu\nu} + B(t^2,r^2,\theta)\, \Gamma^2_{\mu\nu\mu_1\mu_2}. \tag{3.3.47}$$

Again we can flip the charge-conjugation properties with an overall factor of $(r \cdot t)$.

Having derived the system of BSEs describing glueball as well as the general form of the Bethe-Salpeter amplitudes corresponding to a given set of quantum numbers, we will proceed in the next chapters to the application of the work laid out so far. We will numerically study a simple truncation, that preserves all necessary symmetries as well as the high-momentum behaviour of the constituent ghosts and gluons known from resummed perturbation theory. In order to do that we will have to solve the coupled set of equations of motion of ghosts and gluon in the complex plane, so that we are able to solve the glueball BSE for negative euclidean momenta, which correspond to timelike minkowskian momenta, where physical bound states can be formed. the solution of the SDEs of ghosts and gluons in the complex momentum plane, will be presented in the next chapter and after that the results obtained for ghosts and gluons are plugged into the glueball BSE, which is then solved.

Chapter 4

Foundations and treatment of the Schwinger-Dyson equations

4.1 The Yang-Mills System

In section 2.3 we have introduced the coupled system of SDEs for quarks (2.3.16), gluons (2.3.14) and ghosts (2.3.15). To obtain a set of coupled and consistent equations for these fields alongside their bound states, we have introduced a derivation scheme starting from the *2PI* effective action (3.1.24), (3.1.25). In that scheme several parts of our truncation scheme have already been included. Especially since the diagrams contributing to the effective action are at most one-loop, the resulting BSEs are in a truncation that can be considered as being a generalised ladder truncation. The system of equations describing quarks, gluons and ghost was found to be

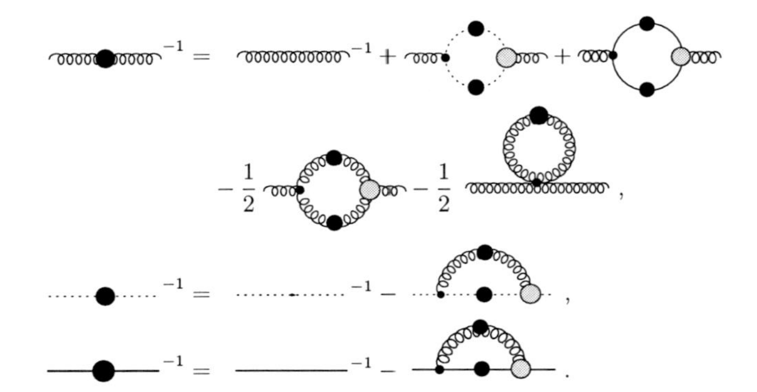

Neglecting the quark and also directly dropping the tadpole diagram in the gluon equation, for it only will contribute an overall normalisation factor, which is renormalised away at the end, we find the set of equations

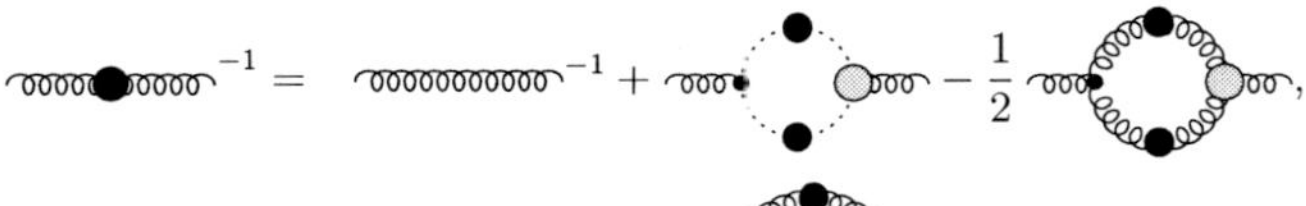

(4.1.1)

(4.1.2)

The shaded blobs indicate dressed vertices. The system (4.1.1) and (4.1.2) is often called the *Yang-Mills system* and has been thouroghly investigated before [HvSA98, vSHA98, Blo01, FA03].

Since the vertex dressings are merely momentum dependent scalar functions multiplied with the corresponding tree-level vertex, they do not change the algebra of the equations. Thus the results of the contractions of tensors in the above equations are independent of the vertex dressing functions and we can resort to the treatment of the Yang-Mills system given in [Fis03]. The first problem that has to be solved, when trying to find numerical solutions to (4.1.1) and (4.1.2) is that there is a quadratic divergence in the first equation.

Written out explicitly the above equations read

$$\left(\delta_{\mu\nu} - \frac{p_\mu p_\nu}{p^2}\right) \frac{p^2}{Z(p^2)} = \left(\delta_{\mu\nu} - \frac{p_\mu p_\nu}{p^2}\right) Z_3 p^2 + \Sigma_{Z,1}(p^2) - \frac{1}{2}\Sigma_{Z,2}(p^2) \tag{4.1.3}$$

$$\frac{p^2}{G(p^2)} = \tilde{Z}_3 p^2 - \Sigma_G(p^2). \tag{4.1.4}$$

Clearly the left-hand side of the first equation is transverse and so has to be the right. However this is only true in exact calculations. The equations will have to be regularised. We will have to resort to a sharp momentum cut-off scheme for the numerical calculations, but that will break transversality of the right-hand side of the gluon equation and thus introduce quadratic divergences. Being created by regularisation, the quadratic divergences are unphysical and can be removed. To identify these terms one can project both sides of the gluon equation (4.1.3) with the pseudoprojector

$$\mathcal{P}_{\mu\nu} = \delta_{\mu\nu} - \zeta \frac{p_\mu p_\nu}{p^2}. \tag{4.1.5}$$

The result will produce quadratically divergent terms on the right-hand side, which are proportional to $(4 - \zeta)$ and unphysical. Those can be simply subtracted[1] [Blo01, Fis03]. Of course having (4.1.3) and (4.1.4) regularised, we need to renormalise to get rid of the cut-off dependence. We will employ a MOM-scheme to do that. We subtract (4.1.3) and (4.1.4) by themselves at fixed renormalisation points μ_Z and μ_G respectively[2]. After the projecting both side with the transverse projector we find

$$\frac{1}{Z(p^2)} = \frac{1}{Z(\mu_Z^2)} + \Sigma_{Z,1}(p^2;\mu_Z^2) - \frac{1}{2}\Sigma_{Z,2}(p^2;\mu_Z^2) \tag{4.1.6}$$

$$\frac{1}{G(p^2)} = \frac{1}{G(\mu_G^2)} - \Sigma_G(p^2;\mu_G^2), \tag{4.1.7}$$

[1] There are also issues coming from the breaking of gauge invariance. However we will not be concerned with those. Being the first calculation of glueballs using a consistent approach of BSEs and DSEs, a deep study of gauge dependences is left for later investigations.

[2] A very convenient side-effect is that the renormalisation factors Z_3 and $\tilde{Z}_3$ drop out. We therefore do not have to calculate them on the fly in each iteration step, when we solve the coupled equations. The renormalisation factors can be calculated afterwards, which will be necessary since they have to be used to calculate the renormalisation factor of the four-gluon vertex which appears in the BSE later on. However the question of multiplicative renormalisability and whether or not a truncation scheme preserves it, is an important matter. But since our scheme does preserve it, as was shown in [Fis03], we will consider that further.

with

$$\Sigma_{Z,1}(p^2;\mu_Z^2) = \Sigma_{Z,1}(p^2) - \Sigma_{Z,1}(\mu_Z^2) \tag{4.1.8}$$

and so forth. Choosing in-going momenta to be positive, we have the renormalised bare vertices

$$\begin{aligned} {}^{(cA)}\hat{V}_{\mu}^{abc}(p_Z;p_{in},p_{out}) &= -\,i\tilde{Z}_1\, f^{abc}\, p_{out,\mu} & (4.1.9)\\ {}^{(A^3)}\hat{V}_{\mu\nu\rho}^{abc}(p_1,p_2,p_3) &= iZ_1\, f^{abc}\left(\delta_{\mu\nu}(p_{1,\rho}-p_{2,\rho})+\delta_{\nu\rho}(p_{2,\mu}-p_{3,\mu})+\delta_{\rho\mu}(p_{3,\nu}-p_{1,\nu})\right). & (4.1.10)\end{aligned}$$

If we choose the dressed vertices to have the general form

$$\begin{aligned} {}^{(cA)}V_{\mu}^{abc}(p_Z;p_{in},p_{out}) &= \frac{1}{\tilde{Z}_1}\,{}^{(cA)}\Gamma(p_Z;p_{in},p_{out})\cdot{}^{(cA)}\hat{V}_{\mu}^{abc}(p_Z;p_{in},p_{out}) & (4.1.11)\\ {}^{(A^3)}V_{\mu\nu\rho}^{abc}(p_1,p_2,p_3) &= \frac{1}{Z_1}\,{}^{(A^3)}\Gamma(p_1,p_2,p_3)\cdot{}^{(A^3)}\hat{V}_{\mu}^{abc}(p_1,p_2,p_3), & (4.1.12)\end{aligned}$$

we can write the self-energies in (4.1.6) and (4.1.7) more explicitly as[3]

$$\begin{aligned} \Sigma_{Z,1}(p^2) &= \frac{g^2}{(2\pi)^4}\int d^4q\, K_{Z,1}(p^2,q^2,(p-q)^2)\cdot\\ &\quad {}^{(cA)}\Gamma(p^2;(p-q)^2,-q^2)\,G(q^2)\,{}^{(cA)}\Gamma(-p^2;q^2,-(p-q)^2)\,G((p-q)^2) & (4.1.13)\\ \Sigma_{Z,2}(p^2) &= \frac{g^2}{(2\pi)^4}\int d^4q\, K_{Z,2}(p^2,q^2,(p-q)^2)\cdot\\ &\quad Z(q^2)\,{}^{(A^3)}\Gamma(q^2,-p^2,-(p-q)^2)\,Z((p-q)^2) & (4.1.14)\\ \Sigma_{G}(p^2) &= \frac{g^2}{(2\pi)^4}\int d^4q\, K_{G}(p^2,q^2,(p-q)^2)\cdot\\ &\quad {}^{(cA)}\Gamma(-q^2;p^2,-(p-q)^2)\,Z(q^2)\,{}^{(cA)}\Gamma(q^2;(p-q)^2,-p^2)\,G((p-q)^2). & (4.1.15)\end{aligned}$$

We see that the only parts missing are the vertex dressings ${}^{(cA)}\Gamma$ and ${}^{(A^3)}\Gamma$ of the ghost-gluon vertex and the three-gluon vertex respectively. In an exact calculation those would have to be determined from their corresponding equations. However this is laborious at best and we will rather use effective vertex dressings, which represent a kind of model building. We will thoroughly account for our choices of the effective vertices in the following.

[3] We will not detail the algebra of equations (4.1.6) and (4.1.7) here. They can be found explicitly in [Fis03].

The ghost-gluon vertex

In principle all correlation functions have their own equations of motion i.e. SDEs. In QCD SDEs of an n-point function typically depend on higher correlations up to $n+2$-point functions, which themselves in turn depend on even higher correlations and so forth. Thus the exact determination of a given n-point correlation entails solving a system of infinitely many coupled SDEs, which today seems rather impossible. It is therefore necessary to rely on approximations, which truncate the tower of coupled equations at some given order. However such a truncation has an impact to the system of equations that is *a priori* uncontrolled. Whether a truncation is suitable or not can only be checked on the result, which can be compared with known information obtained from different approaches e.g. perturbation theory in the range where it is applicable, Slavnov-Taylor identities, symmetries etc. Constructing a truncation thus means to try to include as much knowledge of the behaviour of the physics described by the SDEs as possible, while on the other hand keeping the problem tractable. In the following we will do exactly that. Our truncation scheme acts on all correlation functions higher than propagators. We replace the three-point vertices by sound and tractable constructions and set all higher vertices to zero. We keep the symmetries of the gauge-fixed Lagrangian and include all modifications in appropriate dressing functions, trying to anticipate the impact of the corresponding higher equations of motion by external knowledge of the behaviour of the respective vertex dressings.
For the ghost-gluon vertex we choose the bare form obtained from the renormalised Lagrangian

$$ {}^{(cA)}V_{\mu}^{abc}(p_Z;p_{in},p_{out}) = {}^{(cA)}\hat{V}_{\mu}^{abc}(p_Z;p_{in},p_{out}) \tag{4.1.16} $$

$$ \Rightarrow \qquad {}^{(cA)}\Gamma(p_Z;p_{in},p_{out}) = \tilde{Z}_1. \tag{4.1.17} $$

Furthermore we take

$$ \tilde{Z}_1 = 1. \tag{4.1.18} $$

Thus we use the unrenormalised bare ghost-gluon vertex. Even though taken at face value this might look as a rather rough approximation, it is remarkably good and has been the starting point for almost all investigations of the SDEs of the Yang-Mills sector of QCD. It is based on an idea of Taylor [Tay71] and has been found to be quite accurate in lattice studies as well as in functional investigations [CMM08, SMWA05, AFLE05].

The three-gluon vertex

The three-gluon dressing is a bit more complicated. We will determine a possible dressing by three physically reasonable requirements

1. The dressing shall be such that the resulting dressing functions for ghost and gluons reproduce the high-momentum behaviour known from perturbation theory.
2. The dressing shall preserve Bose-symmetry.
3. The dressing shall not interfere with the expected power-law solution known from analytical studies in the low-momentum sector.

For the first requirement we look at the high-momentum behaviour of equations (4.1.1) and (4.1.2). It is known from resummed perturbation theory that the dressing functions of ghosts and gluons behave like[4] [GW73b, Wei60]

$$G(p^2) = G(\mu^2)\left(1+\omega\log\left(\frac{p^2}{\mu^2}\right)\right)^{\delta} \tag{4.1.19}$$

$$Z(p^2) = Z(\mu^2)\left(1+\omega\log\left(\frac{p^2}{\mu^2}\right)\right)^{\gamma} \tag{4.1.20}$$

for large euclidean momenta. Here we have $G(\mu^2)$ and $Z(\mu^2)$ being the values at some large euclidean renormalisation point μ^2 of the ghost and the gluon dressing function respectively. The exponents δ and γ are the anomalous dimensions of the dressing functions that can be determined order by order in resummed perturbation theory. Suppressing quark contributions to one-loop order one finds [GW73b]

$$\delta = -\,9/44 \tag{4.1.21}$$

$$\gamma = -\,13/22. \tag{4.1.22}$$

[4] A detailed and pedagogical treatment can be found in [Sch08].

The constant $\omega = 11\, N_C\, \alpha(\mu^2)/12\pi$ itself depends on the running coupling $\alpha(p^2)$ at the renormalisation point μ^2, but is of no interest in the following. To meet the first requirement, we have to study the UV behaviour of equation (4.1.1). It is known that the ghost equation is subleading in the UV, so we concentrate on the gluon equation. The interplay of the diagrams in the gluon equation has been studied in [FA03]. There it is found that all diagrams run conformly in the UV, so we can furthermore concentrate on the construction of the three-gluon vertex dressing. We assume that the vertex dressing only depends on squared momenta. Written out explicitly we have

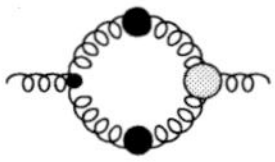

$$= c \cdot \int_0^\infty \int_0^\pi dy\, y \sin^2\theta \, \frac{K(x,y,z)}{xy}\, Z(y)\, Z(z)\, {}^{(A^3)}\Gamma(x,y,z). \tag{4.1.23}$$

Here we have already performed the integral over two angles which are trivial and set $p^2 = x$, $q^2 = y$ and $(p-q)^2 = z$. Furthermore we have gathered all numerical constants into a prefactor c. The integral kernel $K(x,y,z)$ reads

$$\begin{aligned} K(x,y,z) =& \frac{1}{8xy} z^2 + \left(\frac{1}{x} + \frac{1}{y}\right) z - \frac{9}{4}\left(\frac{x}{y} + yx\right) - 4 + \left(\frac{x^2}{y} - 4(x+y) + \frac{y^2}{x}\right) z^{-1} \\ &+ \left(\frac{x^3}{8y} + x^2 - \frac{9xy}{4} + y^2 + \frac{y^3}{8x}\right) z^{-2}. \end{aligned} \tag{4.1.24}$$

We now apply an angular approximation, where we neglect the momentum p^2 in the logarithms. This is justified since these vary slowly for large momenta and the region where z becomes small (i.e. the approximation less precise) is quite narrow. After this approximation only the kernel contains the angle θ. Since we have $z = x + y - 2\sqrt{xy}\cos\theta$, we see that the angular integral in (4.1.23) can be performed using

$$\int_0^\pi \frac{\sin^{2\mu-1}\theta\, d\theta}{(1 + 2a\cos\theta + a^2)^\nu} = B\left(\mu, \frac{1}{2}\right) F\left(\nu, \nu - \mu + \frac{1}{2}; \mu + \frac{1}{2}; a^2\right), \tag{4.1.25}$$

with $\mathrm{Re}\,\mu > 0$ and $|a| < 1$.

In order to do this we split the momentum integral into two parts from $[0 : x]$ and $[x : \infty]$ and factor out $z_1 = x(1 + y/x - 2\sqrt{y/x}\cos\theta)$ and $z_2 = y(1 + x/y - 2\sqrt{x/y}\cos\theta)$ respectively. We find that the hypergeometric function gives nothing but a polynomial in x/y or y/x, except for the terms proportional to z^{-2} where the result is a rational function. Therefore the result of the angular integral can be obtained by applying the following replacements to the integral kernel (4.1.24).

$[0:x]$	$[x:\infty]$
$1/z^2 \to \frac{1}{x^2-xy}$	$1/z^2 \to \frac{1}{y^2-xy}$
$1/z \to \frac{1}{x}$	$1/z \to \frac{1}{y}$
$z \to x+y$	$z \to x+y$
$z^2 \to x^2+3xy+y^2$	$z^2 \to x^2+3xy+y^2$

We thus find

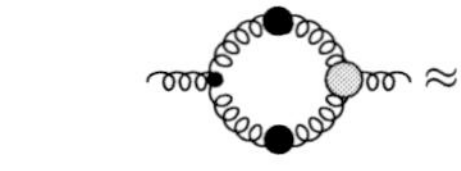

$$\approx \frac{c\pi}{2} Z^2(\mu^2) \cdot \int_0^x \frac{dy}{x} \left(1+\omega \log\left(\frac{y}{\mu^2}\right)\right)^{\gamma} {}^{(A^3)}\Gamma(x,y) \cdot \left(\frac{7y^2-50xy-36x^2}{8x^2}\right) + \frac{c\pi}{2} Z^2(\mu^2) \cdot \int_x^\infty \frac{dy}{x} \left(1+\omega \log\left(\frac{y}{\mu^2}\right)\right)^{\gamma} {}^{(A^3)}\Gamma(x,y) \cdot \left(\frac{7x^2-50xy-36y^2}{8y^2}\right). \tag{4.1.26}$$

The second requirement is that the Ansatz for the three-gluon vertex should be Bose-symmetric. If we choose

$${}^{(A^3)}\Gamma(x,y,z) = a\,\mathfrak{Re}\left[\left(1+\omega\log\left(\frac{x}{\mu^2}\right)\right)^{-\gamma} \left(1+\omega\log\left(\frac{y}{\mu^2}\right)\right)^{-\gamma} \left(1+\omega\log\left(\frac{z}{\mu^2}\right)\right)^{-\gamma}\right], \tag{4.1.27}$$

plug it into equation (4.1.26) and remember that we applied the angular approximation we find that the dressing function contributicns of the gluons are canceled leaving[5]

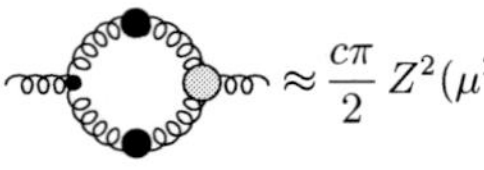

$$\approx \frac{c\pi}{2} Z^2(\mu^2) \left(1+\omega\log\left(\frac{x}{\mu^2}\right)\right)^{-\gamma} \cdot \int_0^x \frac{dy}{x} \cdot \left(\frac{7y^2-50xy-36x^2}{8x^2}\right) + \frac{c\pi}{2} Z^2(\mu^2) \left(1+\omega\log\left(\frac{x}{\mu^2}\right)\right)^{-\gamma} \cdot \int_x^\infty \frac{dy}{x} \cdot \left(\frac{7x^2-50xy-36y^2}{8y^2}\right). \tag{4.1.28}$$

[5] We assume that x is sufficiently larger than the scale μ^2, so the logarithms are positive.

This integral is divergent and we will regularise it by a subtraction method like it is done in the full numerical calculation. We find

$$\Big|_{x} - \Big|_{\xi} = \frac{c\pi}{2} Z^2(\mu^2) \left(1 + \omega \log\left(\frac{x}{\mu^2}\right)\right)^{-\gamma} \cdot \left(-\frac{3\xi^3}{2x^3} - \frac{25\xi^2}{8x^2} + \frac{7\xi}{8x} - \frac{43}{12}\right). \quad (4.1.29)$$

which is dominated by the last term for large x. We thus find that indeed the choice (4.1.27) consistently satisfies equation (4.1.6). The third requirement is that the three-gluon vertex dressing does not interfere with the power-law behaviour of the solution to (4.1.6) and (4.1.7). We will later take a deeper look into the low-momentum behaviour of these equations. For the time being we only need to know that for low momenta equation (4.1.6) is dominated by the ghost-loop unless we spoil that with an unsuitable three-gluon vertex dressing and that the gluon dressing function vanishes like a power for low momenta, while the ghost dressing function either diverges like a power or becomes constant. It is therefore clear that our choice (4.1.27) for the three-gluon vertex does not interfere with the infrared solutions, since it only diverges logarithmically and thus leaves the gluon-loop subleading in the infrared.

With these constructions all parts of the equations (4.1.6) and (4.1.7) are known. What is left are the boundary conditions under which the system is to be solved. These are intimately connected to the infrared behaviour of the solutions, which will now be discussed.

Two types of solutions: scaling and decoupling

In order to calculate the solution of the system (4.1.6) and (4.1.7) there are several parameters left that have to be fixed. Firstly there are parameters stemming from the renormalisation procedure. We used a MOM-scheme which entails that we subtract the equations at some fixed momenta, which have to be chosen. In the case of the gluon equation (4.1.6) it is useful to choose a large momentum μ_Z^2 to do the subtraction. In fact we will subtract the equation at the ultraviolet cutoff. Furthermore we have to choose a value for the gluon dressing function at that subtraction point. This choice is arbitrary and we will use $Z(\mu_Z^2) = 0.5$. For the ghost equation matters are more

subtle. It has been shown that there exist two types of solutions for the system of gluon and ghost SDEs [LvS02, FA03, BLLY$^+$08]. Technically speaking the two types differ by the low-momentum behaviour of the ghost dressing function. In one case the dressing function becomes constant for vanishing momentum. This type of solution is called the *decoupling* solution. The alternative is that the ghost function diverges like a power in the infrared, which is denoted as the *scaling* solution. In both cases the dressing function of the gluon vanishes in the infrared. In case of decoupling it does so proportional to the squared momentum. For scaling the gluon dressing function vanishes like a power of the squared momentum with an characteristic exponent, which is larger than one i.e. it vanishes faster like the squared momentum. This in turn means that the gluon propagator vanishes in the infrared for the scaling solution, while it is constant for decoupling. A very important property of the scaling solution is that the exponents of the ghost dressing function and of the gluon dressing function are not independent [AvS01].
It is possible to choose the type of solution which will be obtained by subtracting the ghost equation at a low momentum. In case of the decoupling solution we choose the infrared cutoff as subtraction point and a finite value for the ghost function at that point not much larger than order one. In fact we will choose $G(\mu_G^2) = 10$. To obtain the scaling solution we have to take the subtraction point μ_G^2 to be very small, almost zero. The requierement to be satisfied strictly would be $1/G(0) = 0$, which is not possible to implement exactly. However doing numerics we can choose an approximation to that requirement using the subtraction point $\mu_G^2 = 10^{-30}$ and $G(\mu_G^2) = 10^{30}$. This is sufficient to generate the scaling solution.

However matters are much more intricate beyond pure numerical technology. The boundary condition for the ghost function in the infrared is intimately related to gauge-fixing.

There has been quite some debate of late on the matter of existence of the two solutions and which one is physical. The first discovered was the scaling solution [LvS02, FA03]. Later on it was shown that a set of power-law behaved ansätze for the infrared behaviour of the correlation functions self-consistently solves the whole untruncated tower of SDEs for low-momenta [FP07, FP09]. A virtue of this solution type is that it is consistent with the picture of colour confinement by Kugo and Ojima [KO79, Kug95] as well as a prediction made by Gribov concerning the impact of Gribov-copies on the dressing function of the ghost, namely that it diverges for low momenta, in the presence of Gribov-copies[6] [Gri78]. As mentioned in section 2.1 simply integrating over all field configurations when calculating the QCD-partition function, includes field configurations connected to each other by gauge-transformations, although from each set of such connected configurations only exactly one should be taken into account. To remedy that Gribov suggested to restrict the domain of integration on the first Gribov horizon (2.1.24). He furthermore predicted that such a restriction leads to a characteristic behaviour of the ghost and the gluon dressing function in the infrared

$$Z(p^2) \propto \frac{p^4}{p^4 + M^4}, \qquad G(p^2) \propto \frac{M^2}{p^2} \tag{4.1.30}$$

for vanishing p^2. M is called *Gribov mass*. More thorough investigations using SDEs showed that the behaviour of the dressing functions in the infrared in Landau-gauge has the form of power laws

$$Z(p^2) = A \cdot \left(p^2\right)^{\rho}, \qquad G(p^2) = B \cdot \left(p^2\right)^{\sigma}. \tag{4.1.31}$$

There is a scaling relation found between the exponents

$$\rho + 2\sigma = 0, \tag{4.1.32}$$

which gives rise to a non-trivial infrared fixed point of the running coupling, when it is defined non-perturbatively as

$$\alpha(p^2) = \alpha(\mu^2) Z(p^2, \mu^2) G^2(p^2, \mu^2), \tag{4.1.33}$$

which has been taken to be renormalised at μ^2. Because of (4.1.31) effectively there is only one free exponent to choose, which most often is denoted with κ. One finds

$$Z(p^2) = A \cdot \left(p^2\right)^{2\kappa}, \qquad G(p^2) = B \cdot \left(p^2\right)^{-\kappa} \tag{4.1.34}$$

[6] A detailed treatment of the infrared behaviour of the scaling scenario in Yang-Mills theory can be found in [AvS01].

and

$$1/2 < \kappa < 1, \tag{4.1.35}$$

while in Gribovs investigation he found $\kappa = 1$. The infrared fixed point of the running coupling (4.1.33) is found to be

$$\alpha(p^2 \to 0) = \alpha(\mu^2) A B^2. \tag{4.1.36}$$

This solution is called the scaling solution [LvS02, FA03]. Additionally this type of solution is in accord with the confinement-scenario of Kugo and Ojima [KO79, Kug95]. There the authors claim that colour confinement in Landau gauge is realized when the dressing function of the ghost diverges for vanishing momentum. They find that there is a relation between the ghost dressing function and an auxiliary function u at zero momentum

$$G(0) = (1 + u(0))^{-1}, \quad \text{with} \quad u(0) = -1, \tag{4.1.37}$$

so the ghost dressing function diverges. The realisation of the Kugo-Ojima confinement scenario along with the power-law behaviour predicted by Gribov made this solution very attractive from the physical point of view. However it was discovered that there exist another type of solutions [BLLY+08], where it is found that in equation (4.1.31) one has the exponents

$$\rho = 1, \qquad \sigma = 0 \tag{4.1.38}$$

and a running coupling that vanishes for low momenta. Even though this solution seems not to reproduce the results of Gribov and Kugo and Ojima considerable support came from the studies of ghost and gluon propagators on large lattices, which seem to reproduces the decoupling solution. Kondo argues that this possibility of two solution types arises from gauge-fixing [Kon09]. He claims that the confinement condition Kugo and Ojima found is not exact. Instead the condition should be

$$G(0) = (1 + u(0) + w(0))^{-1} \tag{4.1.39}$$

with an additional function $w(p^2)$. He then calculates the functions $u(p^2)$ and $w(p^2)$ in *Gribov – Zwanziger theory*, where the Gribov horizon is explicitly included on the

Lagrangian level [Zwa89]. He finds that the Gribov horizon after all determines the value of the ghost dressing function at zero. So parts of the gauge-fixing are implemented by choosing the boundary conditions of the Yang-Mills system. This allows for both types of solutions and entails that they arise from different fixing of some remnant gauge freedom left over by the usual gauge-fixing procedure. In this framework (Gribov-Zwanziger theory) the dressing function of ghosts and gluons also have been investigated. Again both types of solutions have been found [HAS10] and the same is true studying the Yang-Mills system in the version that is derived from the Functional Renormalisation Group [PLNvS04].
However lattice studies in gauge-fixed Yang-Mills theory seems to only find dressing functions of ghosts and gluons corresponding to the decoupling solution [SIMP$^+$06, CM10, OS09, IMPS$^+$07]. The reason for that is most likely gauge-fixing as was pointed out in [Maa10b, Maa10a]. In fact since fixing to Landau-gauge is done statistically in typical lattice calculations, there might be a possibly important difference between the gauge fixed on the lattice and the one from continuum theory.

Since there is plenty of evidence that both types of solutions do exist, in this work we will examine both possible types of solutions without prejudice. Because the effects of gauge-fixing seem to be important only in the deep infrared we do not expect them to have large impact on the bound states of gluons, since the important scales on which the physics of bound state formation take place are much larger. As already explained we can choose the type of solution using the boundary parameter $G(0)$.

Details and results

The numerical treatment of the Yang-Mills system is rather delicate. Having subtracted the unphysical quadratic divergence in the gluon SDE still the system is logarithmically divergent, which entails that the loop integration has to be done on a rather wide range of momenta. This can be safely done by choosing a logarithmic integration grid. However due to the angular integration the functions $G(p^2)$ and $Z(p^2)$ are probed beyond the cutoffs and we need to provide an extrapolation. For this we rely on the knowledge

of the infrared and ultraviolet behaviour of these functions. So we will employ a power-law behaviour in the infrared and the one-loop logarithmic behaviour known from perturbation theory in the ultraviolet. We will allow the system to find the exponent of the power-law dynamically in the infrared, while fixing the result to the correct UV behaviour and then look for converging solutions.
Fixing to the correct logarithm for high-momenta is justified by the ultraviolet analysis from section 4.1. However these restrictions render the system quite stiff.
The solution will be obtained using Newton's method, but still the starting guess has to be quite accurate to obtain a solution to a given set of parameters. Also the range within which the parameters (especially the parameter a) of the three-gluon vertex dressing (4.1.29) can be varied is rather narrow. Furthermore the three-gluon dressing introduce a scale into the numerical calculation.
It is an important feature of the Yang-Mills system that it generates a scale dynamically. To show this numerically, in [Fis03] the three-gluon vertex was constructed such, that it does not introduces a scale and allows the system to generate one. However since it is not our intention to redo this proof, we deliberately use a three-gluon vertex dressing that introduces a scale. Still in our calculation, a dynamical scale will arise nevertheless, for this is a fundamental property of the Yang-Mills system. Therefore we will have two scales present in our calculation, one artificial, introduced via the three-gluon vertex dressing and one arising dynamically. But the scale generated dynamically will be very large compared to the scale we introduce via the three-gluon vertex dressing. We therefore have the situation that the two scales are well seperated and we do not expect significant mutual influences between the scales.
The general properties of the solution will be dominated by the dynamical scale everywhere but in the high momentum region as desired. In the ultraviolet region the logarithmic running of the three-gluon vertex dressing will determine the behaviour of the solutions of the Yang-Mills system as shown in section 4.1. Still there will be some small interplay between the parameters appearing in the three-gluon vertex dressing and the dynamically generated scale. Yet practically we find that in total there is only one significant scale appearing in the solution. In fact by far the most important scale is the one generated dynamically. The scale introduced into the system via the three-gluon

vertex dressing only determines when the logarithmic running of the three-gluon vertex dressing effectively sets in and the aforementioned small mutual interaction between the three-gluon vertex dressing scale and the dynamical scale only dynamically adjusts the position of the transition between medium and high momentum region.[7]
This scale defines some internal units, in which the calculation is effectively performed. Since it is the only scale appearing, we can transform the result to a physical choice of units. To do so we can either use gauge-fixed lattice calculations of the gluon propagator or fix the scale to reproduce bound state mass spectra obtained from non-gauge-fixed lattice results for glueballs. Here we will choose to adjust the scale to the scale obtained from gauge-fixed lattice calculations[8] [OSIS07].

To be concrete we choose the following parameters.[9] The cutoffs we choose as $\epsilon^2 = 10^{-4}$ and $\Lambda^2 = 10^5$ for infrared and ultraviolet respectively. We subtract the gluon equation at $\mu_Z^2 = \Lambda^2$ and the ghost equation at $\mu_G^2 = \epsilon^2$ for decoupling and $\mu_G^2 = 10^{-30}$ for scaling. We take $Z(\mu_Z^2) = 0.5$ and either $G(\mu_G^2) = 10^{30}$ to obtain the scaling solution or $G(\mu_G^2) = 10$ for the decoupling solution as boundary conditions. The value of $G(\mu_G^2)$ for the decoupling solution depends on the choice of the parameters of the three-gluon vertex, since the numerical stiffness of the system does not allow for arbitrary combinations of both. In general for a given set of parameters of the three-gluon vertex there is a minimum value $G(\mu_G^2)$ has to assume, however any larger value is also possible. With these values fixed and the connection between the three-gluon vertex parameters and the value of $G(\mu_G^2)$ in the decoupling case, there remains only the choice of parameters for the three-gluon vertex. The construction we chose was

$$
{}^{(A^3)}\Gamma(x,y,z) = a\,\mathfrak{Re}\Bigg[\left(1+\omega\log\left(\frac{x}{\mu^2}\right)\right)^{-\gamma} \left(1+\omega\log\left(\frac{y}{\mu^2}\right)\right)^{-\gamma}\left(1+\omega\log\left(\frac{z}{\mu^2}\right)\right)^{-\gamma}\Bigg],
$$

[7] In fact the resulting scale defines what medium and high momentum actually means.

[8] This means we rescale the arising internal units such that the resulting maximum of the gluon dressing function coincides with the corresponding maximum obtained in lattice calculations.

[9] We choose internal units to be dimensionless.

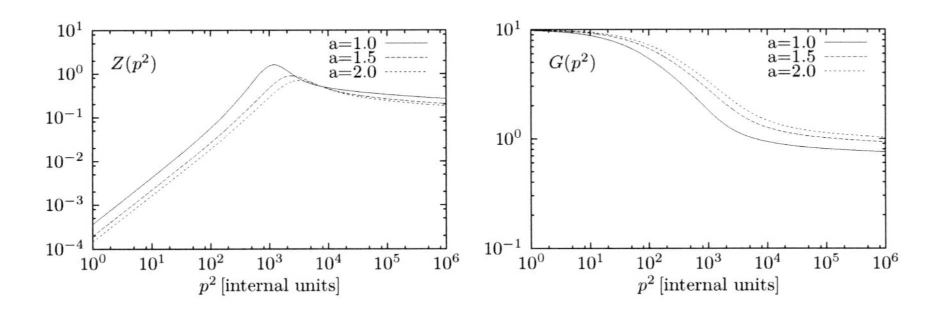

Figure 4.1: Dressing functions of gluons (left) and ghosts (right) for the decoupling solution in internal units.

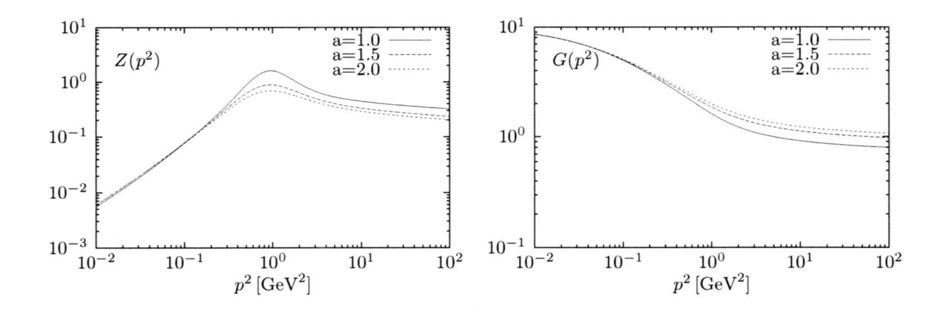

Figure 4.2: Dressing functions of gluons (left) and ghosts (right) for the decoupling solution after rescaling units to lattice results.

Since we have introduced a scale into the system by the choice of the three-gluon vertex dressing,[10] which sets internal units in which we calculate and we will connect these units to physical units afterwards, there is no reason to explicitly modify it by hand, so

[10] The scale is determined by the combination of μ^2, ω and the scale which is created dynamically in the system so it is non-trivial to estimate it beforehand.

we choose $\mu^2 = 1$. We also choose $\omega = 0.1$ for simplicity. The anomalous dimension of the logarithms is fixed by resummed perturbation theory and we take the one-loop approximation, which gives a value $\gamma = -13/22$[GW73a, Wei60].

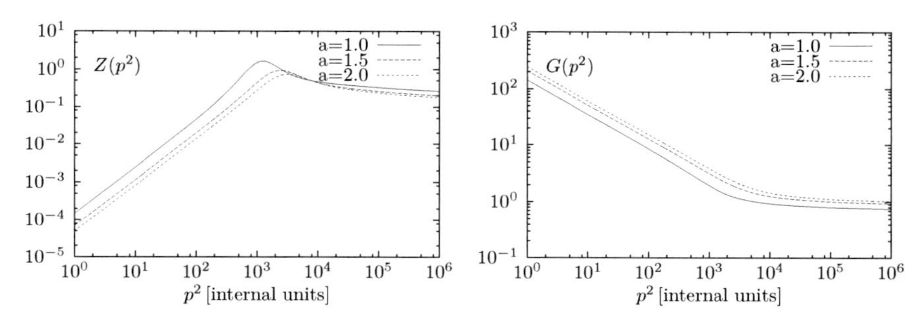

Figure 4.3: Dressing functions of gluons (left) and ghosts (right) for the scaling solution in internal units.

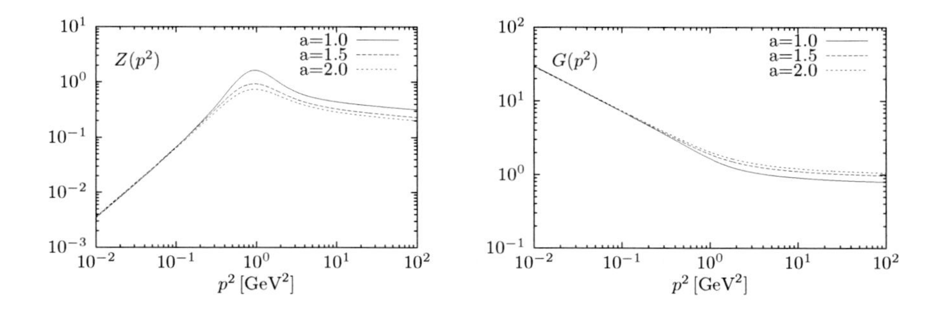

Figure 4.4: Dressing functions of gluons (left) and ghosts (right) for the scaling solution after rescaling units to lattice results.

What remains is the parameter a. As already pointed out the parameter a cannot be chosen arbitrarily because of the numerical stiffness of the system. We will choose a in the range $a = [1.0 : 2.0]$. It will turn out that the system will generally find solutions which have a rather large numerical value of the emerging scale, which however is only slightly sensitive to the choice of a. Rather than moving the scale about, the parameter a modifies the shape of the gluon dressing function. This is not surprising, since the behaviour of the gluon dressing function for high and low momenta is already known and fixed. Therefore only the intermediate momentum range, in which the solution turns from the perturbative to the low momentum behaviour is determined dynamically, so it is clear that the variation of the parameters mostly will influence the way, how such a turnover will exactly look like.

Having made our choices of parameters, we can finally calculate the gluon and ghost dressing functions. In figures 4.1 to 4.4 we show resulting solutions for the dressing functions of ghosts and gluons in both the decoupling and the scaling case. We show them in internal units as obtained from the numerical calculation and after rescaling to physical units by means of lattice results [OSIS07, BHL+07, IMPS+07].

The infrared exponent κ (cf. eq.(4.1.34)) in the scaling case is found to be $\kappa \simeq 0.6$, which is in good agreement with the value $\kappa = 0.595353$ determined in [FA03]. In figure 4.1 and 4.3 we see that the variation of a while keeping the other parameters fixed leads to a variation of shape and scale of the resulting dressing functions. While the shape of the ghost dressing function does not change much, the gluon dressing function changes in the intermediate momentum region. The bump in the gluon function becomes more or less pronounced, which in the glueball BSE will lead to different integrated strengths of interaction in the momentum region, which is most important for the formation of bound states of gluons. However after rescaling to physical momentum scales we see in figures 4.2 and 4.4 that the change of scales is removed, so the variation of a effectively only changes the shape of the dressing functions.

4.2 The Yang-Mills system for complex momenta

4.2.1 Foundations

As will be seen in section 5.2 to determine masses of physical bound states from a BSE, it is necessary to solve it for negative total squared momenta if working in euclidean spacetime. Clearly this means that the total momentum becomes complex. Thus in the loop-integrals of the BSE there will appear complex momenta. Especially in the combined approach using SDEs to generate dressing functions of the constituents as input for the BSE, there will arise the necessity to calculate these dressing functions for complex momenta which means that it is necessary to solve the SDEs for such complex values. This has been done in several SDE/BSE studies before e.g. in [MT00, MR03, FW08, AWW02]. The strategy adopted in most cases is to work in ladder truncation. In the SDE the momentum routing is then chosen such that the constituent particles of the BSE are always purely real. Then a model *ansatz* is chosen for the exchange particle of the BSE appearing in the respective SDE. The model is chosen such that it is easily calculated for complex momenta and then the SDE is solved. To clarify the method consider the combined system of quark SDE and meson BSE for identical quark flavours

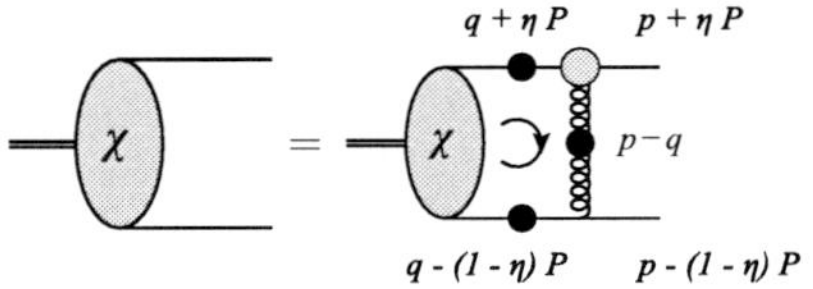

χ = χ ; $q+\eta P$, $p+\eta P$, $p-q$, $q-(1-\eta)P$, $p-(1-\eta)P$ (4.2.1)

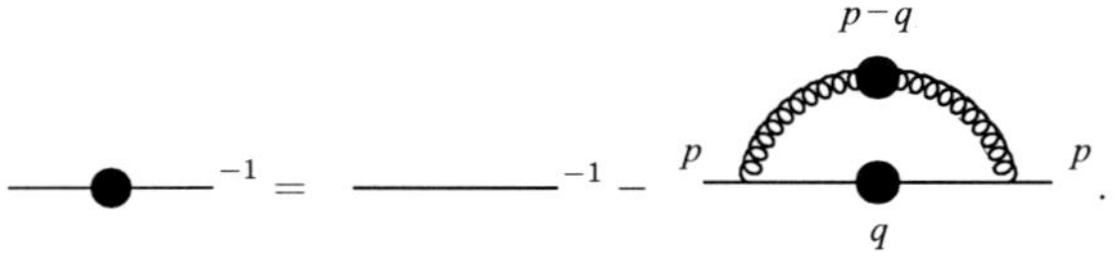

$^{-1}$ = $^{-1}$ − ; $p-q$, p, p, q . (4.2.2)

We have chosen the momentum p to be complex in the SDE. In the BSE the total momentum P is purely imaginary so the dressing functions of the quarks have to be known for complex momenta and the exchange gluon stays real, while it becomes complex in the SDE. The advantage is that the gluon can be chosen such that it is known in the complex plane *a priori*. Then the corresponding dressing functions of the constituents i.e. quarks are simply solutions of (4.2.2) for complex values of the gluon momentum. A different approach was chosen only recently in studies of the meson BSE beyond ladder truncation [11][FW08, FW09, FNW09]. In these studies it was not possible to keep the quark in the SDE having purely real momenta and a way of solving the SDE for explicitly complex quark momenta, has been presented in [FNW09]. This method exploits the fact that the dressing functions of the quarks depend only on squared momenta and that the BSE, for which we calculate the dressing functions as input, is solved for fixed total momenta. This leads to the observation that the integrals in the BSE probe constituent momenta that lie on parabolae in the complex plane. The crucial observation is that the vertex of those parabolae depends only on the total momentum. Furthermore these parabolae are determined by only two parameters and we thus find a very useful property of the momenta probed by the loop-integral of the BSE. Having chosen a fixed total momentum for which we want to solve the BSE, we can pick a momentum point appearing in the integral. The vertex, fixed by the total momentum, and that point determine a parabola in the complex momentum plane. We want to determine the value of the quark dressing functions at that particular point by solving the SDE (4.2.2). To do this we need information about values of the dressing functions for the various loop-momenta appearing in the integral of the SDE. Since we are only interested in solutions of the SDE for quark momenta that will also appear in the BSE we choose $p_{SDE} = q_{BSE} + \eta P$ in (4.2.2), for a particular momentum point that we need in the BSE.[12] However the only free parameter left in the SDE then is the loop momentum q, which is purely real. Thus all complex momenta probed by the loop integral in the SDE, will lie on the same parabola as p_{SDE} if the momenta are oriented parallel. If they are not, we find that only the interior of the parabola will be probed. That allows for a convenient algorithm to solve the SDE for complex momenta as needed in the BSE. One first solves the SDE for positive real momenta. Then a set of parabolae in the complex plane is chosen, each of which having a vertex shifted further along the negative real axis thus enveloping its predecessors. Now one solves the SDE successively on the parabolae stepping further outward and using the

[11] More precisely beyond rainbow-ladder truncation, since most prominently the quark-gluon vertex was replaced by a more sophisticated *ansatz*.

[12] The solution for complex conjugate momenta, we take as the conjugate of the solutions for the original momenta e.g. $A(\bar{p}^2) = \bar{A}(p^2)$.

information of the solutions on the previous parabolae by interpolating between them. Let us consider the algorithm in a more formal way. We define

$$q_+^2 := (q + \eta P)^2 \tag{4.2.3}$$

$$q_-^2 := (q - (1-\eta)P)^2. \tag{4.2.4}$$

The total momentum P is purely imaginary around some neighborhood of a physical bound state. We can write (4.2.3) and (4.2.4) in terms of absolute values

$$q_+^2 = q^2 - \eta^2 P^2 - 2i\eta q P \cos\theta \tag{4.2.5}$$

$$q_-^2 = q^2 - (1-\eta)^2 P^2 + 2i(1-\eta) q P \cos\theta, \tag{4.2.6}$$

where θ is the angle between the momentum four-vectors q and P. We set

$$x_\pm \equiv \mathfrak{Re}(q_\pm^2) \tag{4.2.7}$$

$$y_\pm \equiv \mathfrak{Im}(q_\pm^2) \tag{4.2.8}$$

and eliminate the absolute value q in (4.2.7) using (4.2.8). We find the parabola in the complex plane on which all momenta with fixed P and variable loop-momentum (now in terms of x) and θ lie[13]

$$y_+ = 2\sqrt{x + \eta^2 P^2}\,\eta P \cos\theta \tag{4.2.9}$$

$$y_- = -2\sqrt{x + (1-\eta)^2 P^2}\,(1-\eta) P \cos\theta. \tag{4.2.10}$$

Considering only y_+ for the moment and setting θ to zero, for some fixed P_0 we find a parabola that envelopes all parabolae with $P < P_0$ if η is also kept fixed. In Fig 4.5 we show a set of such parabolae along with a similar set with $\theta = \pi$ that lies in the lower half-plane. Together these give full parabolae in the complex plane that will be probed in the integral of the BSE. In Fig 4.6 we show the effect that varying θ has on the parabola y_+ when P and η are kept fixed. The resulting parabolae all lie in the area inside the parabolae corresponding to $\theta = 0$ and $\theta = \pi$. Thus given particular P, q and η the angular integration in the BSE will only need information from the inside of the parabola corresponding to these values. The most important effect of the momentum

[13] We choose the sign of the root such that y_+ lies in the upper and y_- in the lower half-plane for $\theta = 0$.

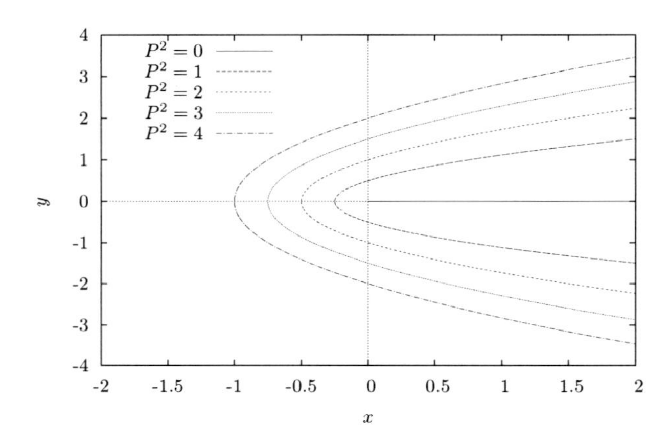

Figure 4.5: Momentum parabolae y_+ for some total momenta with fixed $\eta = 0.5$ and variable q. The parabolae in the upper half-plane correspond to $\theta = 0$, while those in the lower half-plane to $\theta = \pi$.

partition parameter η is that it shifts the vertex of the parabolae y_+ and y_- against each other. Only for $\eta = 0.5$ the (half-) parabolae correspond to the same vertex. In Fig 4.7 a set of such parabolae is shown for various η. This shift is important especially in cases when the bound state is comprised of two particles of different kinds as is explained in section 5.2.

From these considerations we can see that indeed the integration path that will be probed in the BSE is fixed to one parabola if the total momentum P and the momentum partition parameter η are fixed. The reason is that if we choose a particular external momentum p in (4.2.1) we immediately fix a parabola, say y_+ from (4.2.9). If we vary the momentum p we will move along this parabola. However this is exactly what we do when we perform the integral in the BSE (4.2.1). Since we fix the form of the complex momentum $p-q$ in (4.2.2) to resemble the form of the constituent momenta in (4.2.1), we can solve the SDE (4.2.2) step by step on the parabolae. We end up with the following algorithm.

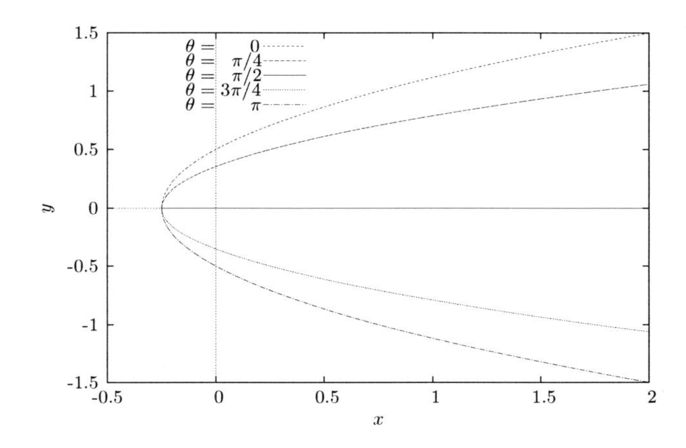

Figure 4.6: Momentum parabolae y_+ for fixed $P = 1$, $\eta = 0.5$ and variable q and θ.

1. Solve the SDE (4.2.2) for positive real squared momenta.
2. Check whether y_+ or y_- from (4.2.9) or (4.2.10) respectively will extend deeper into the negative real axis.
3. Choose a set of squared total momenta on the negative real axis. The total momenta are thus purely imaginary. This set has to extend deep enough into the negative real axis to cover the momentum range or area, that will be probed in the integration of the BSE (4.2.1).
4. Set up the momentum parabolae in the complex plane for $\theta = 0$ and $\theta = \pi$. They are determined by the grid of discrete external momenta p.
5. Calculate a starting guess on the innermost parabola (next to the positive real axis). This can be done by continuing the solution on the real axis obtained in step 1, using the Cauchy-Riemann equations.

6. Find the solution on the first parabola by iteration or using Newton's method. The loop-integral will probe momenta inside the parabola because of the angular integration. To provide the values of the dressing functions we solve for, we employ an interpolation inside the parabola using the information of the dressing function on the parabola and the solution on the real axis. Clearly this interpolation has to be done in each iteration step, since the values of the dressing functions on the parabolae will change.

7. Proceed to the next parabola which encloses all parabolae on which the solution of (4.2.2) has already been calculated. Set a starting guess and calculate the solution with an iteration or with Newton's method. Perform an interpolation for momenta inside the parabola using the solutions on the parabolae obtained in the previous steps and the current guess on the present parabola.

8. Repeat the previous step until the solution of (4.2.2) has been found on all parabolae set up in step 3.

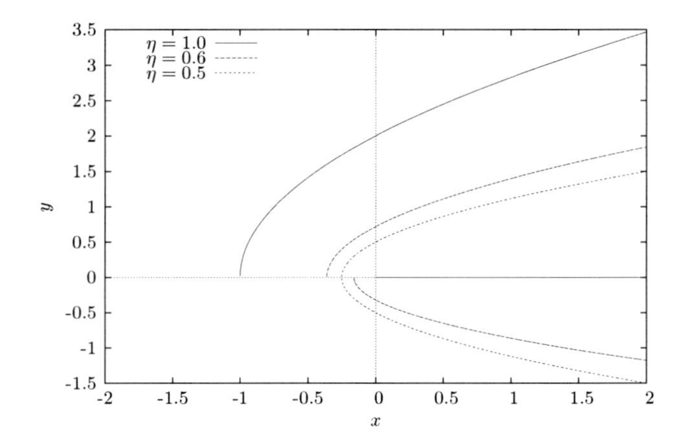

Figure 4.7: Momentum parabolae y_+ and y_- for fixed $P = 1$, $\theta = 0$ and variable q and η. The vertices of the parabolae are shifted against each other except for $\eta = 0.5$.

In the case of the Yang-Mills system this method (called *shell method* [FNW09]) has a very useful property. Recall the Yang-Mills system from section 4.1

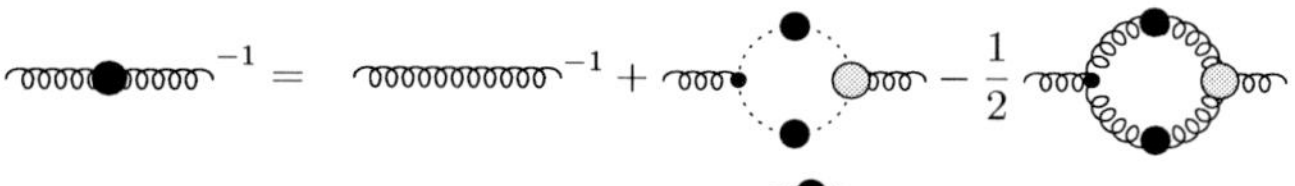

$$(4.2.11)$$

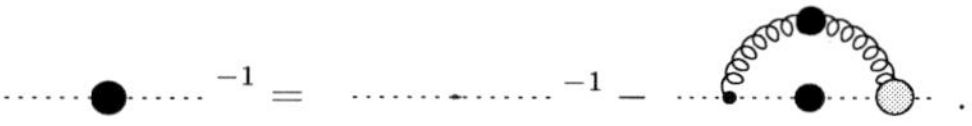

$$(4.2.12)$$

Consider first the ghost equation. We choose the following momentum routing

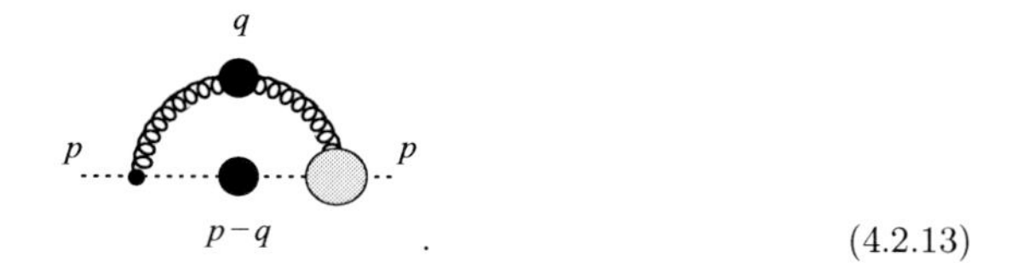

$$. \qquad (4.2.13)$$

The gluon function will only be probed on the positive real momentum axis. Thus having calculated the solution of the system (4.2.11) and (4.2.12) for positive real momenta and choosing the momentum routing (4.2.13) the ghost equation is decoupled from the gluon equation in the complex plane. Using the shell-method we can calculate first the ghost dressing function and use it to solve for the gluon dressing function.

In the next section we will present numerical results for the dressing functions of ghosts and gluons in the complex plane. We will discuss the analytic structure and effects of numerical errors in detail. After that we will consider two analytical fits to those dressing functions that have been obtained in an earlier work. We will use both for the glueball BSE in order to get information about the impact of numerical accuracy on the result of the BSE.

4.2.2 Ghosts and Gluons in the complex momentum plane

In this section we will present numerical results obtained by applying the shell-method described in the previous section to the Yang-Mills system. First we will discuss technicalities and obstacles of the numerical calculation in some detail. Especially the observation that with our method the ghost and gluon dressing functions of scaling type cannot be obtained in the complex momentum plane. Afterwards we will present the solutions of the Yang-Mills system in the complex momentum plane in the case of decoupling. We will then proceed to discuss a set of fit-functions for ghost and gluon dressing functions that can be investigated in the complex momentum plane and is available both for the decoupling and the scaling case. We will then compare the numerical results and the fit-functions.

Details of the numerical calculation

To solve the Yang-Mills system numerically for complex momenta, we will use the shell-method as discussed in section 4.2.1. However we will add another refinement to the shell-method. As discussed previously the key idea for solving the Yang-Mills system was that we first solve for positive real momenta and use the solution obtained, to decouple the ghost and gluon equations. As second step we solve the two equations separately in the complex plane. In order to guarantee the solution obtained so be consistent, we perform an iterative refinement as a third step. We calculate the solutions for ghosts and gluons in the complex plane using solution obtained in the previous iteration until convergence. An important fact is that we cannot obtain both types of solution so far (i.e. scaling and decoupling type cf.. section 4.1). Only the decoupling type solution converges in the complex plane. The problem with the scaling solution seems to be of numerical nature.
In the ghost equation, we do not get convergence on the first shell i.e. the shell that envelopes only the real positive momentum axis (cf.. section 4.2.1). We can stop the iteration on this shell at some arbitrary step and continue calculating the solution on the following shells. These calculation then do converge.
However inspecting the complete result the solution in the complex plane does not look

consistent. The results on the positive real axis looks disconnected from the rest. There is a sharp jump when going from the first shell to the real axis, regardless of how fine the grid formed by the shells is chosen. Interestingly the first shell itself seems consistent with the other shells, although it did not converge. Most likely the reason is that it did not converge because of the inconsistence of the result on the first shell with the positive real axis in its interior, which is probed due to the angular integrations.
The reason for these problems most likely stems from the infrared behaviour of the solution for the ghost function on the positive real axis. The non-trivial power-law behaviour for small momenta, presents a rather intricate numerical difficulty for positive real momenta already. We can incorporate this behaviour when we solve the Yang-Mills system for real momenta and in fact have to, because of the angular integration that always probes momenta beyond the cut-offs. However for complex momenta of low absolute values, we do not have such knowledge to build in. So we have to put some guess into the calculation.[14]
The most obvious choice is to assume the infrared behaviour known for real momenta be valid for small absolute values of complex momenta also. We thus used a constant extrapolation for decoupling type solutions, which works very well. However in case of the scaling solution, neither constant nor power-law infrared behaviour leads to a converging solution in the complex plane.
Assuming the power-law we "almost" get convergence in the sense that the maximal relative error of the ghost function in the complex plane goes done to roughly one or two percent in the first continuation step. Taking that continuation and iterating it together with the corresponding gluon result, we find that both functions stay "almost" the same for a couple of iteration steps and the solutions obtained display a distinctive analytic structure. However we find that continuing the iterations both dressing functions start to blow up.

[14] We can of course simply take the interpolator employed beyond its data range. But that is rather risky, even when using cubic splines. In fact we have done this, but we found that this induces some kind of erratic influence on the analytic structures in the negative half-plane. Anyway it is in general not recommendable to use interpolators outside the data range.

Since the decoupling solution presents itself as very stable and nicely converging, one might think that there is no inherent numerical problem with the equations themselves, but rather some kind of round-off error problem arising from the large values the ghost dressing function assumes for small absolute values of complex momenta. However this is but a guess, for we did not investigate the numerical breakdown of the calculation for the scaling solution in detail.
Since we assume that the deep infrared behaviour of the dressing functions does not have much influence (if any) on glueballs, we will focus on the decoupling solution and furthermore consider two analytic fits to the solution of the gluon dressing function for real momenta. These fits will provide information on the impact of the solution type as well as on the general properties of solutions of the glueball BSE, which will be presented in chapter 5.

Considering the dressing functions of ghosts and gluons for complex momenta in the case of decoupling type solutions we find that the results are mostly smooth and free of numerical noise for momenta in the half-plane of positive real momenta.[15] Only when the imaginary parts of the momenta become very large we get noise, partly due to numerical round-off errors. The other source for these errors are remnant inaccuracies back-feeding from the negative half-plane. Still the smoothness is definitely sufficient as input in the BSE.
The most interesting parts of our results are the dressing functions of ghosts and gluons in the negative half-plane.[16] There we find non-trivial analytic structures and this region is the most important for bound states of gluons.
Also it is numerically somewhat challenging. To generate non-trivial analytic structures numerically without having a starting guess, that more or less resembles those structures from the beginning, is a somewhat intricate task. Calculating the solution step-by-step

[15] The parameter a in the three-gluon vertex dressing is chosen to be $a = 1.0$ throughout the section, since this choice will also be made for the calculations of glueball masses using the BSE in section 5.3. Varying this parameter does not lead to any qualitative changes in the results. Only the scales are modified.

[16] In the following we will refer to the half-plane of positive real momenta (or right half-plane) as the positive half-plane and the negative-half plane vice versa.

starting from the positive real axis, a pole for instance arising at some point will introduce round-off errors into the following steps while rising. Of course in any iterative calculation, changes made in every step will back-feed into following steps, however in case of poles or even branch-cuts these changes become large and their impact can possibly be much larger than the function itself in the following steps, especially in our case, where the next step means next shell and clearly the function on this shell can be small compared to the integrals effected by a rising structure, that eventually becomes a pole on a previous shell.

To deal with these problems and to guarantee that the solution obtained is not only some kind of first step continuation into the complex plane, but instead the true converged solution of the SDEs, we take the solution from the shell method for each dressing function (which are decoupled when taking the solution on the real axis to start off c.f. section 4.2.1) plug them back into the system of SDEs and iterate until convergence.[17]

In general it is important to check, whether arising analytic structures are stable against variation of sample points, used for the calculation.

We did such variations and found that this is not true in general in our calculation. However this is not a problem. We find that there are several pole-like structures on the negative real axis for both ghost and gluon functions and for the real part as well as for the imaginary. But if we increase the precision of the calculation all these structures, except the most pronounced diminish significantly. Of course having only limited time we had to choose a compromise of accuracy and run-time. Still the results are precise enough and the remnants of the spurious structures can only be seen in Figure 4.26 and to some extend in the figures 4.15, 4.18, 4.25 and 4.28. Apart from these spurious structures we find that the results are quite stable against variation of the number of sample points if only we take sufficiently many.

In the following we will show and discuss a variety of figures, which display the numerical results.

[17] The reason why such an iteration of the whole system leads to improving results is that the starting guesses for the dressing functions on the shells improve in each step. Using an iterative method to solve for the dressing functions on each shell, our method is quite sensitive to the quality of the starting guess.

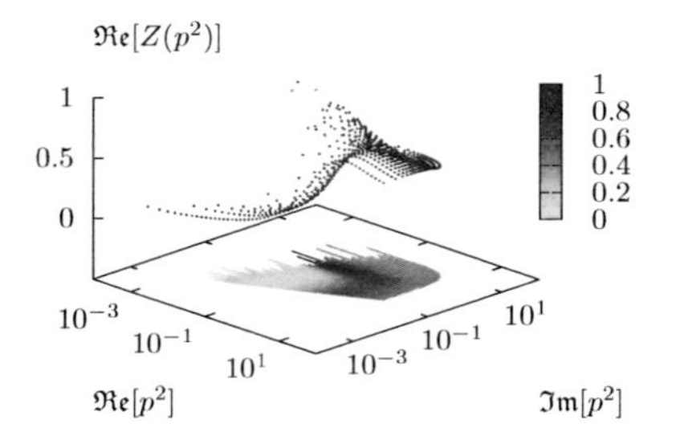

Figure 4.8: The real part of the gluon dressing function in the positive half-plane. Perspective 1.

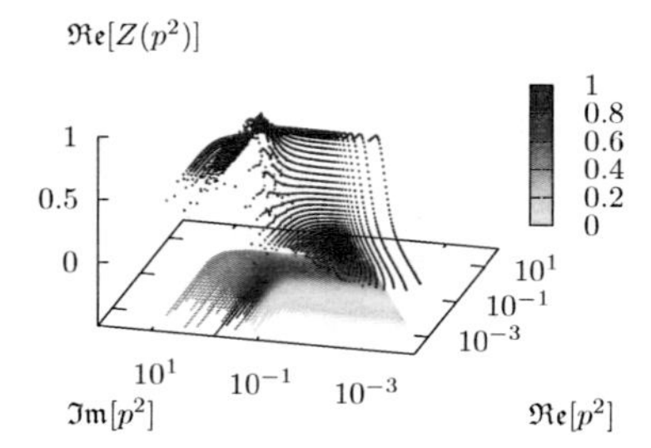

Figure 4.9: The real part of the gluon dressing function in the positive half-plane. Perspective 2.

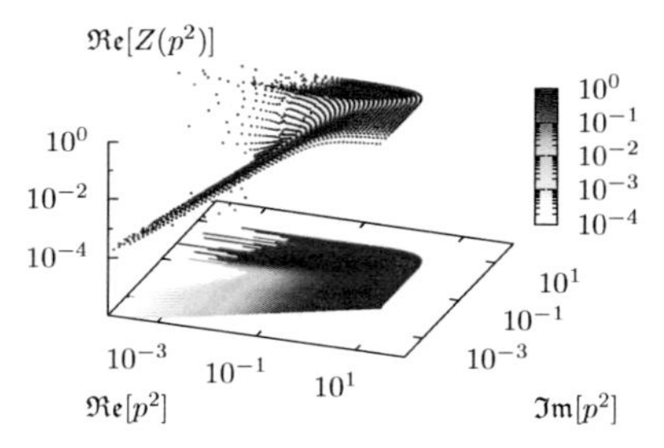

Figure 4.10: The real part of the gluon dressing function in the positive half-plane. The function values are shown on a logarithmic axis to emphasize the power-law behaviour in the infrared.

Numerical results for decoupling solutions

We start with the real part of the gluon dressing function in the half-plane of positive real momenta. Figures 4.8 and 4.9 show the results on a logarithmic scale for both the real and imaginary part of the momentum. The reason is simply that the range of momenta

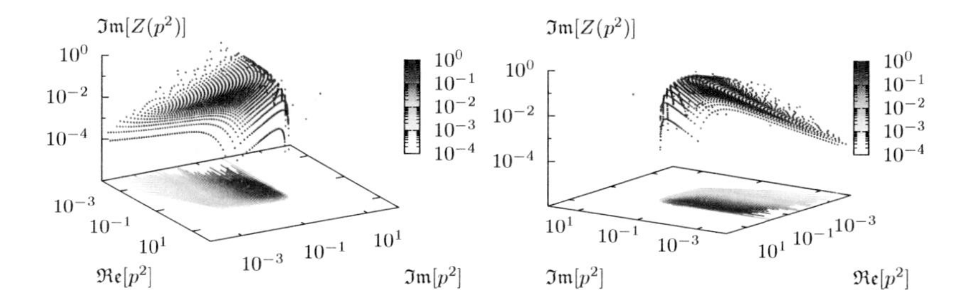

Figure 4.11: The imaginary part of the gluon dressing function in the positive half-plane. Perspective 1.

Figure 4.12: The imaginary part of the gluon dressing function in the positive half-plane. Perspective 2.

is rather large covering several orders of magnitude. Note that all figures showing the numerical results, which are presented in the following are already in physical units obtained by fixing the arising scale-conversion factor (cf.. section 4.1) to the scale found in lattice simulations [OSIS07, BHL+07, IMPS+07].

It is remarkable that the complex dressing functions of ghosts and gluons can be probed rather deeply into the complex plane so well numerically. Even though we will see that the results in the negative half-plane are qualitatively worse, they are still stable and sufficiently smooth.

In figure 4.10 we have shown the gluon dressing function again, this time in triple logarithmic form. Here we can see very well the power-law behaviour for low momenta. In order to provide a sufficient impression of the results in the complex plane we will always show each dressing function at least in two different (mostly more or less opposite) perspectives.

The following figures 4.11 and 4.12 show the imaginary part of the gluon function in the positive half-plane. Again we observe that the result is quite smooth with numerical inaccuracies occurring only for momenta with large imaginary parts. Since the values of the imaginary part of the gluon dressing function are somewhat small for a large range of

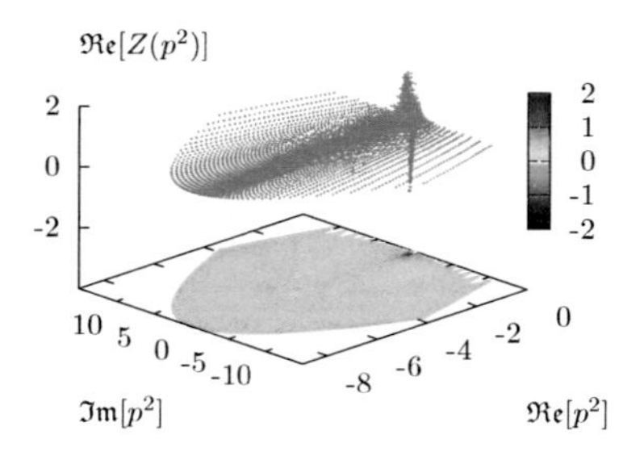

Figure 4.13: The real part of the gluon dressing function in the negative half-plane. Perspective 1. There is a structure at about $0.25\,\mathrm{GeV}^2$ which might be a pole.

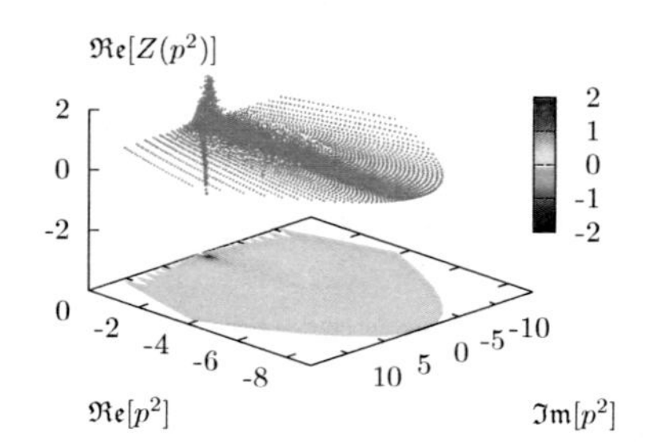

Figure 4.14: The real part of the gluon dressing function in the negative half-plane. Perspective 2.

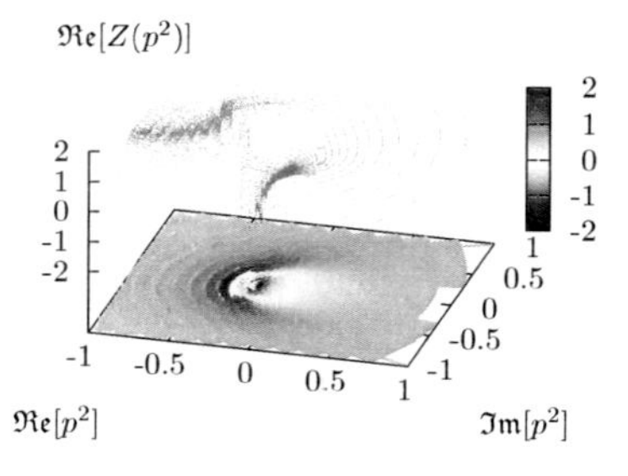

Figure 4.15: The real part of the gluon dressing function around the origin of the complex plane. The pole-like structure is clearly seen.

momenta we chose triple-logarithmic plots. Even though for intermediate momenta it might look as if there was some kind of power-law behaviour, a closer look does not confirm that and we emphasize that such a behaviour is not the reason for choosing the triple-logarithmic plot.

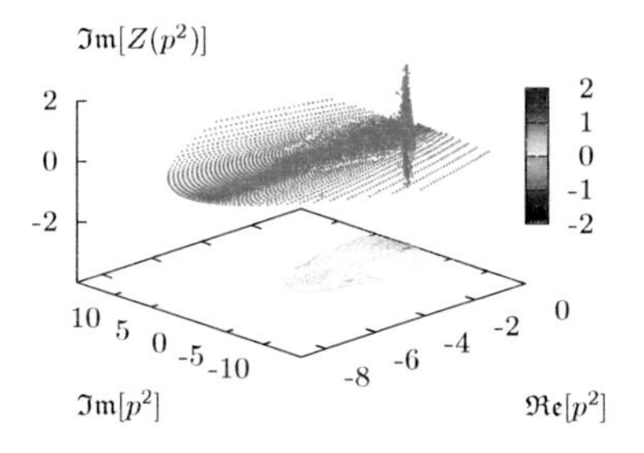

Figure 4.16: The imaginary part of the gluon dressing function in the negative half-plane. Perspective 1. There is a structure at about $0.25\,\mathrm{GeV}^2$ which might be a pole.

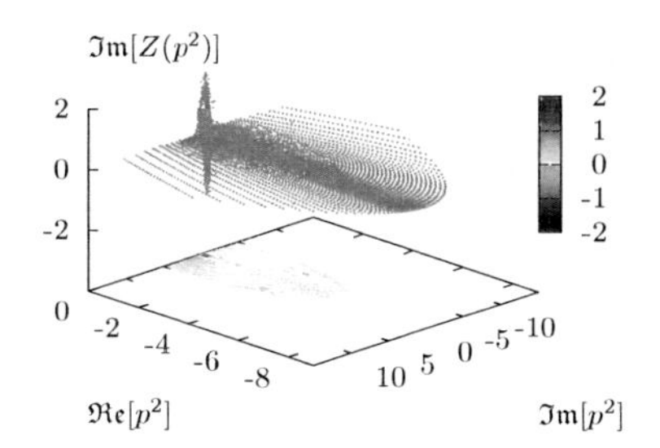

Figure 4.17: The imaginary part of the gluon dressing function in the negative half-plane. Perspective 2.

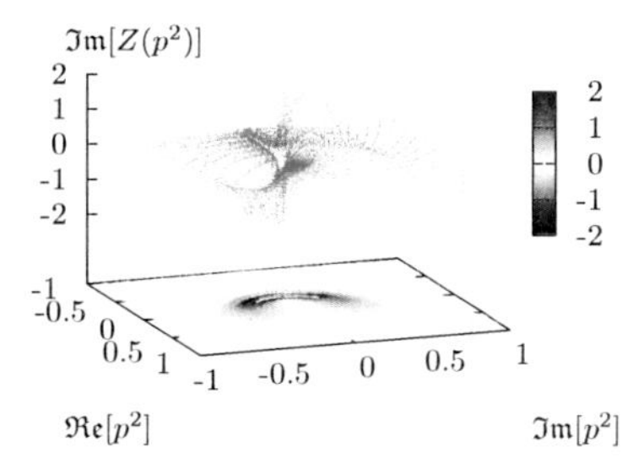

Figure 4.18: The imaginary part of the gluon dressing function around the origin of the complex plane. The pole-like structure is clearly seen.

We will now turn to the negative-half plane starting with the real part of the gluon dressing function. In figures 4.19 and 4.20 we show the result in the negative half-plane. We can clearly see a structure, that seems to be a pole.[18] It is located on the negative

[18] We will not make definite statements about the type of an analytic-structure we encounter. The

real axis at a momentum of about 0.25 GeV^2, which then corresponds to Λ_{QCD}. Apart from this the real part of the gluon function is quite flat, vanishing if we go deeper into the negative half-plane.
We take a closer look in figure 4.15, where we plot the gluon function around the origin. Here we also see the numerical impact of the arising structure to the result on higher shells. Extending into the negative half-plane, we see some kind of ripples in the gluon function. Looking at these ripples in more detail, especially at reduced accuracy, one finds some pattern that reminds on diffractive patterns in wave-mechanics. With increasing accuracy these ripples diminish. Looking closely one can also see the remnant spurious structures on the negative real axis. At some different view point they would be visible more clearly, but we did not want to spent space adding a figure showing that, since as already mentioned, also the spurious structures diminish with increasing accuracy.

Similarly to the real part, the imaginary part of the gluon function is mostly flat and vanishing for large absolute values of the momentum. We find an analytic structure dominating the imaginary part of the gluon dressing function located at the same scale as the dominating analytic structure that we found in the real part of the gluon dressing function. Numerical inaccuracies prevent us from making a suggestion, what type of structure it is, we see. It can be a pole, which is even parallel to the real axis and odd parallel to the imaginary or a pair of complex conjugate poles close to each other. Though the latter possibility seems less likely, when looking at figure 4.18, where the imaginary part of the gluon dressing function around the origin has been plotted.
Having presented the results obtained for the gluon dressing function for complex momenta, we now turn to the respective findings for the ghost. The real part of the ghost dressing function in the positive half-plane is shown in figures 4.19 and 4.20. We can nicely see that it is constant in the infrared not only for small real momenta but also for small absolute values of the momentum. Again we see that for momenta with large imaginary parts slight numerical inaccuracies begin to show up.

reason is that we did not strive to prove a structure to be a pole by calculating a residue, since this is numerically challenging and up to this point not our foremost concern.

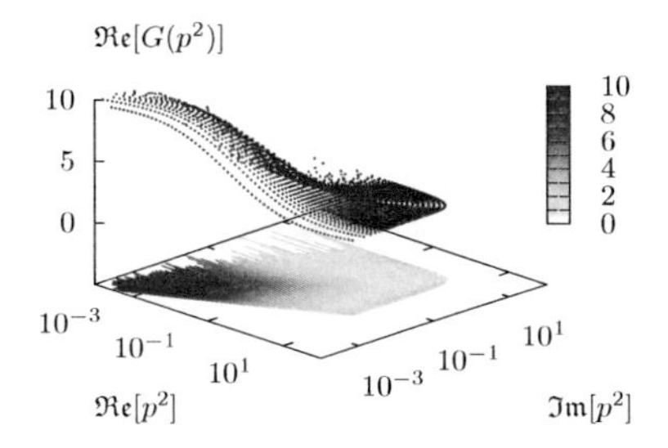

Figure 4.19: The real part of the ghost dressing function in the positive half-plane. Perspective 1.

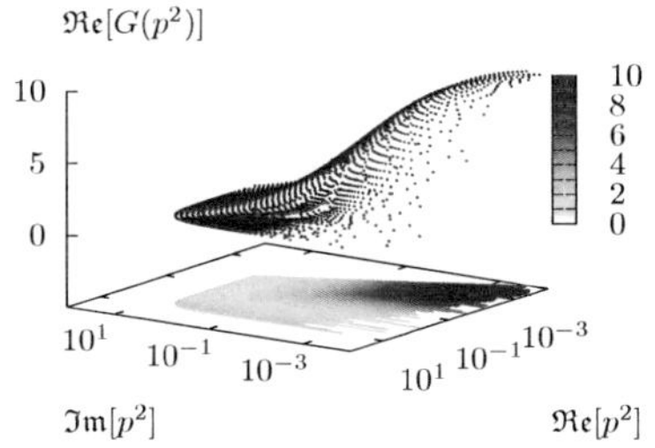

Figure 4.20: The real part of the ghost dressing function in the positive half-plane. Perspective 2.

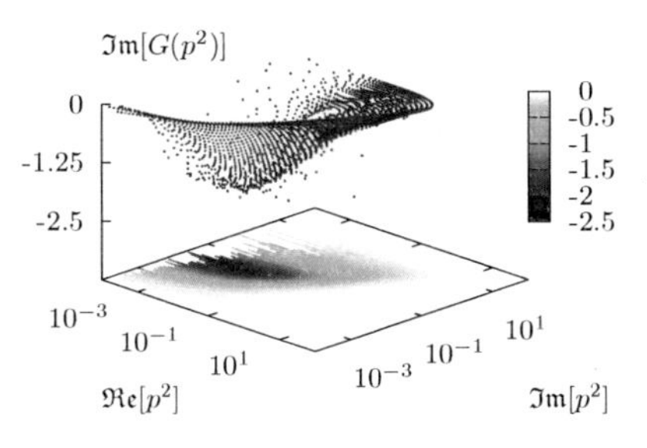

Figure 4.21: The imaginary part of the ghost dressing function in the positive half-plane. Perspective 1.

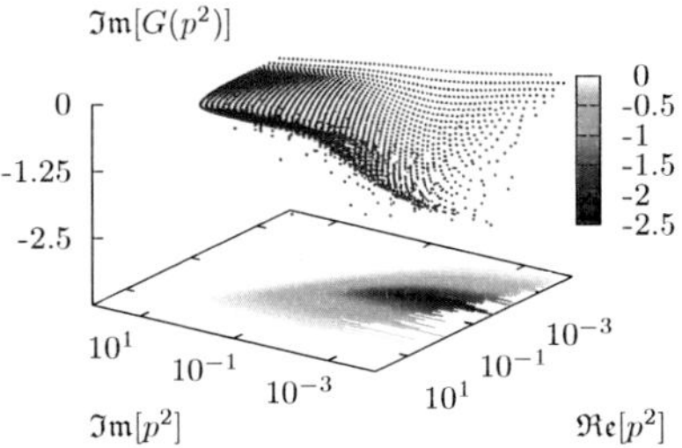

Figure 4.22: The imaginary part of the ghost dressing function in the positive half-plane. Perspective 2.

Figures 4.21 and 4.22 show the imaginary part of the ghost dressing function, which are rather flat and again quite smooth. In the negative half-plane, we first consider the real part. In figures 4.23 and 4.24, we see it is mostly flat and smooth with a sharp structure at the origin. However the logarithmic plots of the positive half-plane in figures 4.13 and 4.14 already indicate that this is not an analytic structure, but some finite "lump",

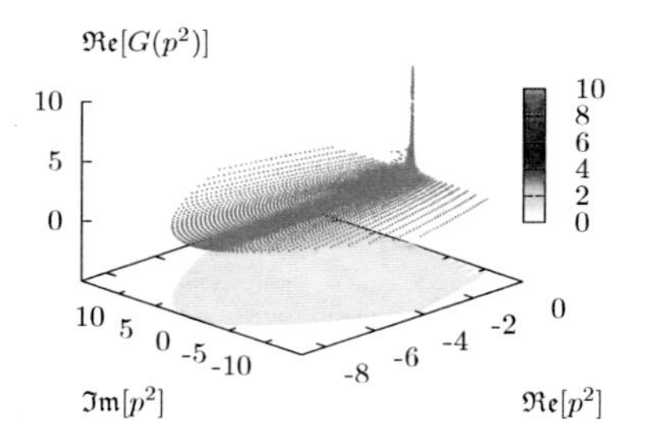

Figure 4.23: The real part of the ghost dressing function in the negative half-plane. Perspective 1. The dominant structure, some kind of "lump", is located at the origin of the complex plane and finite.

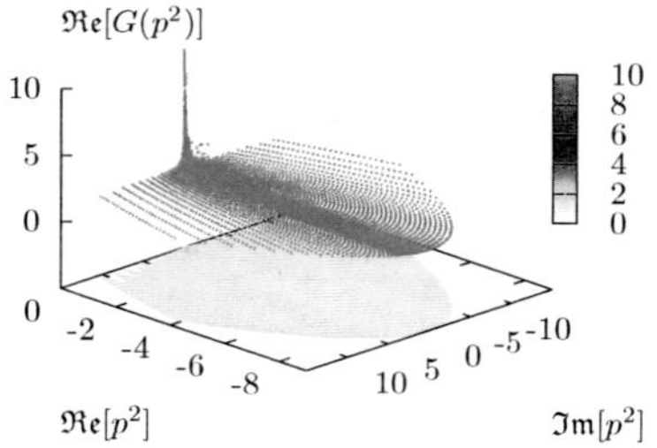

Figure 4.24: The real part of the ghost dressing function in the negative half-plane. Perspective 2.

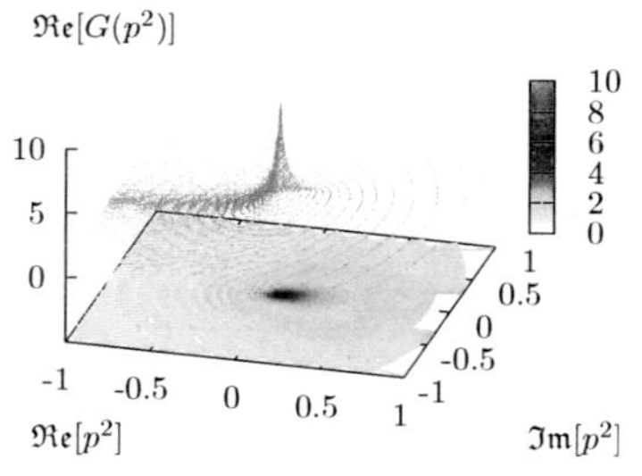

Figure 4.25: The real part of the ghost dressing function around the origin of the complex plane.

which is constant for small absolute values of the momentum. This is confirmed looking more closely at the structure, which is shown in figure 4.25. The effect of the numerical inaccuracies is very nicely seen here. The wave-like ripples and the high density of points around the negative real axis, which is a side effect of the shell-method, along with the

steep rising structure at the origin create an impression of a surface of water shortly after the impact of a drop, where the light falls in from the right and creates a shadow. However there are almost no remnants from the spurious structures along the negative real axis mentioned above.
Finally we have the imaginary part of the ghost function in the negative half-plane plotted in figures 4.26 and 4.27. As before the function is mostly flat. We have a structure at the origin or possibly rather starting at the origin.
At the first glance one might think it a pole being odd parallel to the imaginary axis. However since the real part of the ghost function obviously is finite around the origin, as well as the complex gluon function, it is hard to imagine how such a structure would be generated from the coupled system of equations. More likely the function is also finite and there is a branch-cut opening at the origin, which can be seen quite nicely in figure 4.28. The steep jump which is seen in the plots is due to the numerical treatment.
We also note that we see the spurious structures mentioned earlier. Here they are not yet so small as in the gluon function or the real part of the ghost function, so for the imaginary part of the ghost function we seem to need better accuracy than before.

Having discussed in detail the numerical solution of the coupled system of ghost and gluon SDE, we will now turn to the results of a study conducted earlier. There the authors have generated fit-functions to the numerical results of ghost and gluon dressing functions for real momenta, which can be used in the complex plane. Since in the original paper there was not much emphasize being spent on the properties of the fits for complex momenta, we will do so in the next section and compare the structure of these fits to our numerical results. In chapter 5 we will investigate the glueball BSE and solve it numerically using the numerical results that we presented in the present section as well as the fit-function, we discuss in the next section.

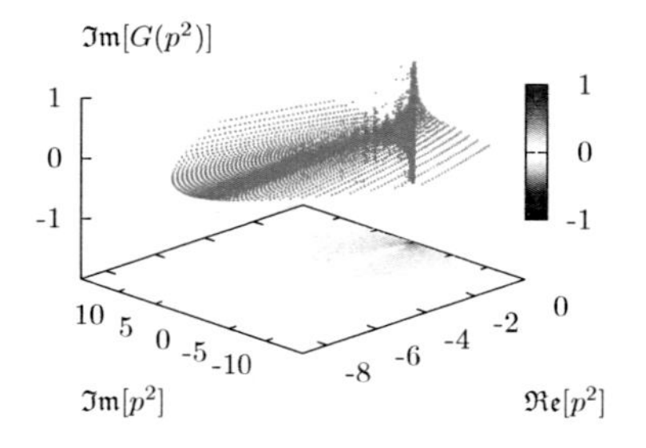

Figure 4.26: The imaginary part of the ghost dressing function in the negative half-plane. Perspective 1. The structure at the origin of the complex plane is finite. It might possibly represent the onset of a branch-cut, which is spoiled by numerics.

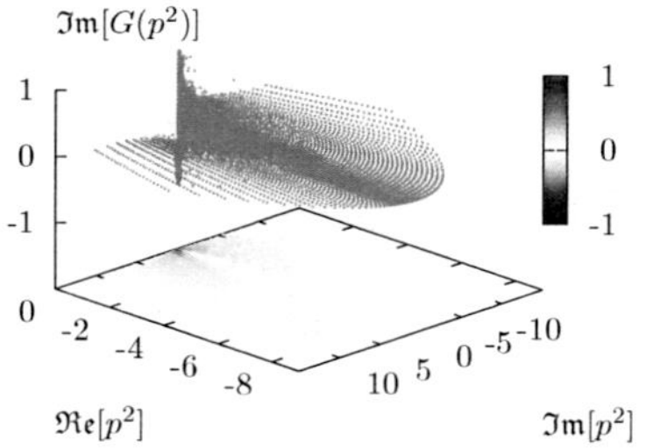

Figure 4.27: The imaginary part of the ghost dressing function in the negative half-plane. Perspective 2.

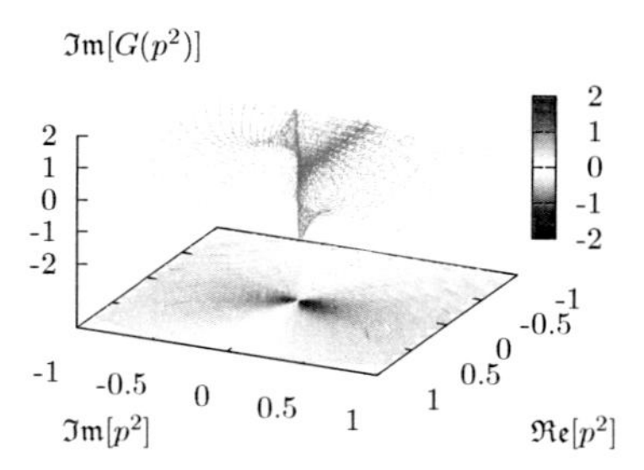

Figure 4.28: The imaginary part of the ghost dressing function around the origin of the complex plane. The very steep transition from upper to the lower half-plane is clearly seen. Possibly there is a branch-cut opening at the origin, which is spoiled by numerics.

4.2.3 Two fits for Ghosts and Gluons

In this section we will discuss fit-functions to the dressing functions of ghosts and gluons. Our starting point is taken from [ADFM04]. There the key idea was to fit an analytic function to numerical data for the scaling type gluon dressing function obtained from the Yang-Mills system for real momenta, as well as to the *Schwinger function* of the gluon. We will shortly recall some terminology and then adapt the work given in [ADFM04] and extend it to obtain fit-functions for ghost and gluon dressing functions for both scaling and decoupling solution.

Constructing fit-functions for ghosts and gluons

The Schwinger function of the gluon is the Fourier transformation of $Z(p^2)/p^2$ over three-space

$$\Delta(t) = \int d^3x \int \frac{d^4p}{(2\pi)^4}\, e^{i(tp_4+\vec{x}\cdot\vec{p})}\, \frac{Z(p^2)}{p^2}.$$

The Schwinger function is a useful tool to study confinement. In quantum field theory the absence of certain states from the observable spectrum can be connected to violation of positivity. That means that expectation values of field-operators get negative contributions from such states. This of course would render the probability interpretation of expectation values useless and thus such states have to be forbidden in the physical spectrum.

The precise mathematical formulation of this criterion has been given by Osterwalder and Schrader for euclidean field theory [OS73, OS75]. It can be formulated in terms of the Schwinger function and basically entails that the Schwinger function be positive to describe a physical field.

The authors in [ADFM04] constructed their fit-function such that they fit both the numerical result of $Z(p^2)$ and of $\Delta(t)$. Doing so they obtained functions valid for complex momenta, whose analytic properties could be studied. In this paper two different fits have been presented, but we will concentrate on the one which fits the data better by eyesight. The combined fits have only been done for the gluon, so we will have to construct a corresponding fit for the ghost ourselves. At the time the fits were presented, only the scaling

solution was known. Thus the fits given by the authors are only of scaling type. We will take these fits and generate decoupling type fits merely by changing the exponents of the infrared part of the fits. The reason is, that the gluon propagator does not change much when switching between scaling and decoupling type of solutions. The difference is pronounced in the ghost function, but the corresponding fit-function we will construct from the gluon fit-function and the running coupling.
The gluon fit-function we will use as given in [ADFM04] reads

$$Z(p^2) = R_Z(p^2)\,\alpha_{fit}^{-\gamma}(p^2), \tag{4.2.14}$$

where we have the fit to the running coupling

$$\alpha_{fit}(p^2) = \frac{\alpha(0)}{1+p^2/\Lambda_{QCD}^2} + \frac{4\pi}{\beta_0}\frac{p^2}{p^2+\Lambda_{QCD}^2}\left(\frac{1}{\log(p^2/\Lambda_{QCD}^2)} + \frac{1}{1-p^2/\Lambda_{QCD}^2}\right) \tag{4.2.15}$$

and an infrared fit

$$R_Z(p^2) = a\left(\frac{p^2}{p^2+\Lambda_{QCD}^2}\right)^{2\kappa}. \tag{4.2.16}$$

The running coupling corresponds to the non-perturbative definition

$$\alpha(p^2) = \alpha(\mu^2)Z(p^2,\mu^2)G^2(p^2,\mu^2), \tag{4.2.17}$$

which has been used in a variety of studies. For scaling it has a non-trivial fixed point at zero momentum. The corresponding value has been calculated in [FA03] and was found to be $\alpha(0) = 8.915/N_c$. The parameters $\beta_0 = (11N_c - 2N_f)/3$ and $\gamma = -13/22$ are known from perturbation theory. Studying the pure gauge-sector of QCD we use $N_c = 3$ and $N_f = 0$. The scale Λ_{QCD} is not fixed to the scale obtained from lattice calculations (which we used to fix the scale of our numerical calculations) at the beginning. Instead we take the value $\Lambda_{QCD} = 420\,\mathrm{MeV}$ used in [ADFM04] and rescale afterwards to the lattice-scale, like we did for our numerical calculation.[19] Before we choose the parameters κ and a, we have to construct the corresponding fit for the ghost dressing function.

[19] The advantage of doing so simply is that we can take the fits and plug them in the numerical calculation of the BSE directly.

We will assume that the fit-function for the ghost has a form similar to the fit-function of the gluon.

$$G(p^2) = R_G(p^2)\, \alpha_{fit}^{-\delta}(p^2), \tag{4.2.18}$$

with

$$R_G(p^2) = b \left(\frac{p^2}{p^2 + \Lambda_{QCD}^2} \right)^{-\lambda}. \tag{4.2.19}$$

Plugging (4.2.14) and (4.2.18) into (4.2.17) and considering first scaling type solutions, for which $\kappa = \lambda$, we directly find

$$-\gamma - 2\delta = 1, \tag{4.2.20}$$

which fixes the anomalous dimension of the ghost to $\delta = -9/44$. Since for the scaling type of solutions we find an infrared fixed point given by $\alpha(0)$ and consistently the limit of our fit-function for the running coupling $\alpha_{fit}(p^2 \to 0) \to \alpha(0)$, we also find $b = 1/\sqrt{a}$. In the scaling case we have $\kappa = 0.595353$. [FA03] The only parameter left is a and we take the choice from [ADFM04] namely $a = 2.7$.

For decoupling, we first note that the running coupling does not have an infrared fixed point, but vanishes for low momenta. Thus our fit (4.2.15) of the running coupling is not valid anymore. However the perturbative behaviour is still right and we choose to simply migrate our analytic forms from the scaling to the decoupling type. In the case of decoupling the gluon dressing function vanishes like a square of the momentum argument, while the ghost dressing function goes to a constant. We thus choose $\kappa = 0.5$ and $\lambda = 0$. Again the parameter a and b are still undetermined. Since in our numerical investigation of the ghost and the gluon function in the complex plane, we chose the decoupling solution with a value for the ghost function at zero momentum $G(0) = 10$, we will do the same here and choose $b = 10/\alpha(0)^{-\gamma}$. The parameter a is chosen such that the result for the numerical gluon function and the fit-function coincide in the infrared after fixing of scales and we find $a \approx 0.354$.

General properties of the fit-functions

Studying the analytic fit-functions to ghost and gluon dressing function gives two benefits. Firstly, we can take the fits corresponding to the decoupling solution and use them in the same way in the glueball BSE as we did with our numerical results. Since this time we will have no numerical inaccuracies, we can study the numerical behaviour of the BSE without any feedback from the numerics of the Yang-Mills system. From that we will get an insight what properties genuinely stem from the BSE and what may come from numerical problems.
Secondly using fits, we are able to study the behaviour of the scaling type solution. Numerically we were not able to generate this type of solution, so having the fits allows us to investigate the impact of the solution type nevertheless. Also comparing the fit-functions and the numerical results is interesting by its own right, the fits being constructed using Schwinger-functions and only data of the ghost and gluon dressing functions for positive real momenta, while the numerical results being the solution of the Yang-Mills system in the complex plane. However there is a difference here. The data from which the fits have been constructed was obtained using a different model-dressing of the three-gluon vertex (see [ADFM04]), so it is not to be expected that the fits and the numerical results are more or less the same. However in fact there are quite some similarities, which indicate that the structure of the solutions of the Yang-Mills system is widely determined by the structure of the equations and that approximations like the model three-gluon dressing do not have too much an impact.
By and large we will see that the numerical result and the fit-function in the positive half-plane are very similar even for both types of fit-functions, which is reassuring. More significant differences will be found in the negative half-plane.[20]

[20] This is not surprising since there are many factors, that can lead to differences between the two results. Firstly one should note, that the decoupling fit-function is an ad-hoc adaption of the one corresponding to the scaling solution. It is left for future investigation, to check how well such a fit actually represents gluon Schwinger-functions for decoupling type of solutions or lattice data. For both types of solution it is true that the fit-function is but an approximation. However, when it comes to analytic structures, even slight differences can give a decidedly different behaviour. This is even more so for the numerical solution, since there are always numerical inaccuracies, which can have an impact. Furthermore analytic structures have to be generated, while solving the set of

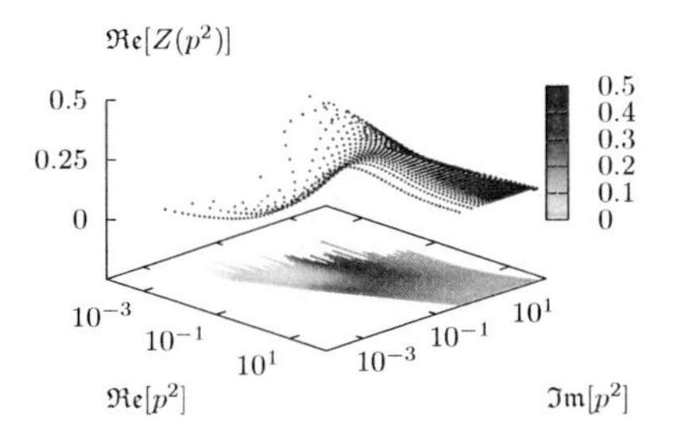

Figure 4.29: The real part of the decoupling type gluon fit-function in the positive half-plane. Perspective 1.

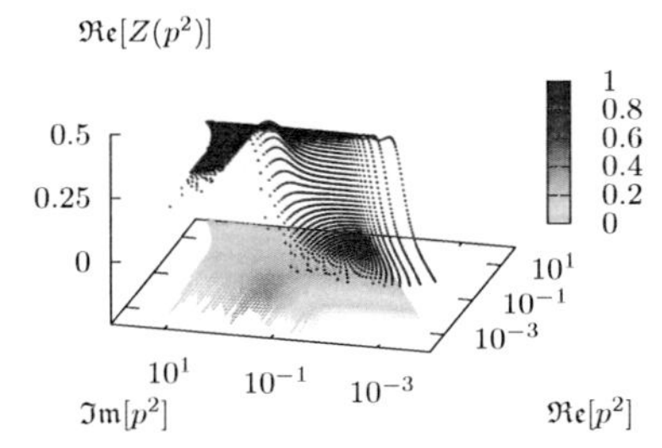

Figure 4.30: The real part of the decoupling type gluon fit-function in the positive half-plane. Perspective 2.

To clarify matters recall eqs. (4.2.14) to (4.2.16). We find that if we set $p^2 = -\Lambda_{QCD}^2$, we get singularities. In both cases these are significant singularities, due to the rational anomalous dimension parameters γ, δ and κ. In all plots of the fit-functions for ghost and gluon dressing function for both types of solutions, in the negative half-plane they can be seen clearly. We have chosen $\Lambda_{QCD} = 0.42\,\text{GeV}$ and after rescaling we find the location of the singularities roughly about $0.3\,\text{GeV}^2$ to $0.4\,\text{GeV}^2$, which is quite close to the position of possible singularities and analytic structures in the respective numerical result, which is located roughly at $0.25\,\text{GeV}^2$. Furthermore the fit-functions have branch-cuts on the negative real axis. In the case of decoupling they stem from the logarithmic ultraviolet part of the fit eq. (4.2.15). For scaling there is another branch-cut coming from the power-law part given by eqs. (4.2.16) and (4.2.19).

SDEs, which might work well or not. All in all this means, that the analytic structure of the fits is not a kind of key requirement we have to meet and we are far from making definite statements about them. This consideration makes the apparent similarity between the numerical results and the fit-function even more remarkable and reassuring that we are doing something right with the numerical treatment as well as with the fit-functions.

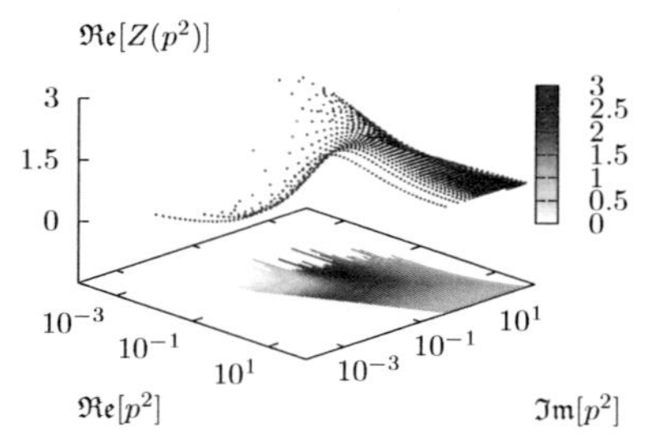

Figure 4.31: The real part of the scaling type gluon fit-function in the positive half-plane. Perspective 1.

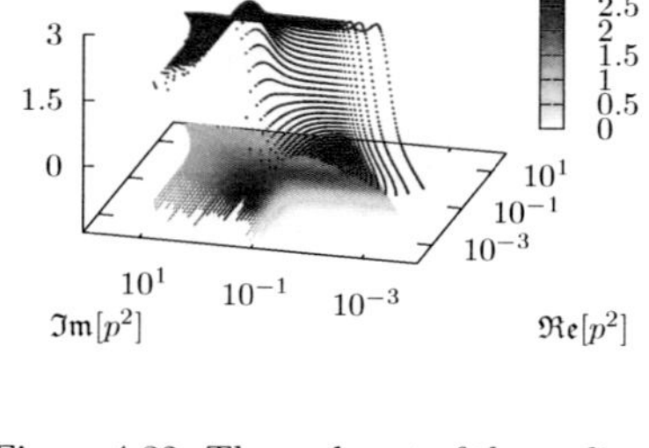

Figure 4.32: The real part of the scaling type gluon fit-function in the positive half-plane. Perspective 2.

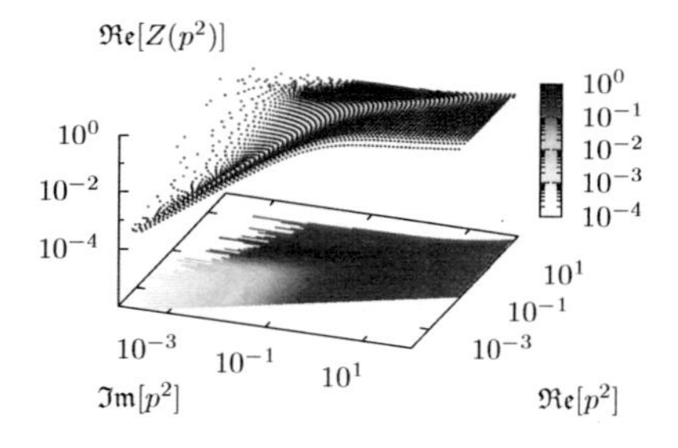

Figure 4.33: The real part of the decoupling type gluon fit-function in the positive half-plane. The function values are shown on a logarithmic axis to emphasize the power-law behaviour in the infrared.

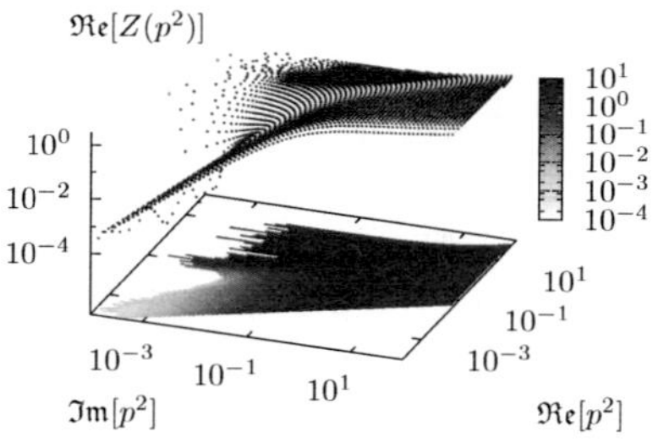

Figure 4.34: The real part of the scaling type gluon fit-function in the positive half-plane. The function values are shown on a logarithmic axis to emphasize the power-law behaviour in the infrared.

On comparison we will see that such branch-cuts are not present in the numerical results.[21] The imaginary parts of the fit-functions will exhibit discontinuities along the branch-cut, as is expected from structure of the respective sources of those cuts.

We will now take a closer look at the results for the fit-functions. We will do so simultaneously for scaling and decoupling solutions, so they can be compared directly.

[21] There are two possible reasons for that. The first is simply numerics. It is unclear, whether a branch-cut could have arisen in the dynamical calculation using the shell method. On the one hand, branch-cuts arise numerical usually as some kind of dynamical fusion of lots of poles or singularities, which are generated dynamically during the iteration process. We already mentioned, that there were lots of singularities during the iteration process in our calculation, but they diminished with increasing accuracy (i.e. number of sample points). Possibly the accuracy was sufficient to suppress the discrete singularities, but still too low to give rise to the generation of a branch-cut. On the other hand, the branch-cuts in the imaginary parts of the fit-functions stem from a purely perturbative result, which naturally is valid in principle only for the high-momentum region. Whether or not an analytic structure, induced by such a perturbative part, still is present for lower momenta in a truly non-perturbative calculation is unclear. It is present in the fit-functions, because they are constructed as a product of a perturbative and a non-perturbative part and even though the perturbative part is subleading for low momenta, its analytic structure is still present. The true non-perturbative result however does not necessarily have to look the same. For instance the "true" solution could be some kind of hypergeometric function, which for large momenta resembles a logarithm, for intermediate momenta behaves like some polynomial and for small momenta exhibits the behaviour of a rational function or a power-law. Then we had a logarithmic behaviour as found in perturbation theory for large momenta and a behaviour as found in purely non-perturbative calculations for small momenta without having a branch-cut coming from the perturbative logarithm on the negative real axis.

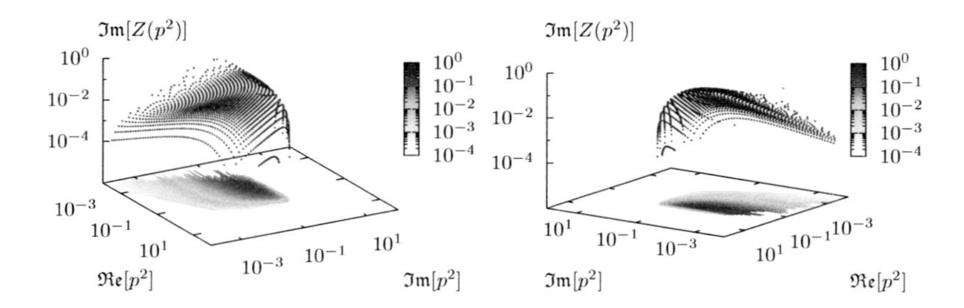

Figure 4.35: The imaginary part of the decoupling type gluon fit-function in the positive half-plane. Perspective 1.

Figure 4.36: The imaginary part of the decoupling type gluon fit-function in the positive half-plane. Perspective 2.

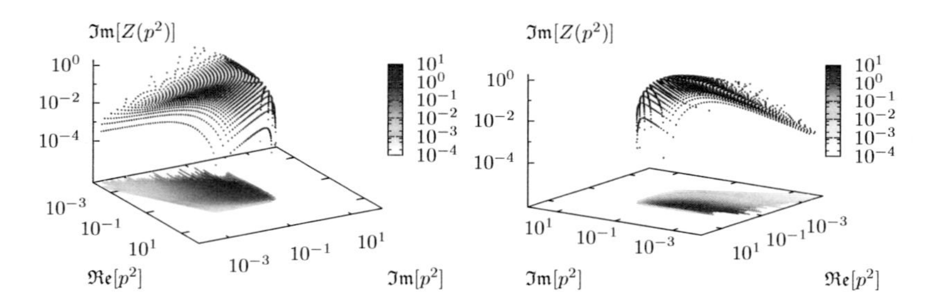

Figure 4.37: The imaginary part of the scaling type gluon fit-function in the positive half-plane. Perspective 1.

Figure 4.38: The imaginary part of the scaling type gluon fit-function in the positive half-plane. Perspective 2.

The gluon fit-functions

The numerical result for the gluon dressing function and the fit-function look remarkably similar in the positive half-plane, both qualitatively and concerning the position of the dominant structure for the real as well as for the imaginary part. This indicates that there is not much difference between the behaviour of scaling and decoupling fit-functions. This is as is expected, for in the positive half-plane the behaviour of both types is dominated by the turning from the infrared part (eqs. (4.2.16) and (4.2.19)) to the ultraviolet-part given in eq. (4.2.15) and the ultraviolet part itself for high momenta. However in the infrared the difference is but the slight difference between the infrared exponents of the power-law (i.e. 0.5 for decoupling and 0.6 for scaling).

In figure 4.29 to 4.32 we show the real part of the gluon fit-function in the positive half-plane for decoupling and scaling respectively. As we did when we presented the numerical results in the positive-half plane, we chose logarithmic momentum axes. Also as before, in the following all momenta are already given in physical units i.e. adjusted to the scale obtained from gauge-fixed lattice calculations [OSIS07, BHL$^+$07, IMPS$^+$07]. In figures 4.33 and 4.34 we used a triple-logarithmic plot to emphasize the power-law behaviour of the gluon fit-functions for momenta with small absolute values. Comparing to the numerical results shown in figures 4.8, 4.9 and 4.10, we find that qualitatively there is hardly a difference. The fit-functions are a bit less steep for intermediate absolute values of momentum. They extend deeper into the region of high absolute values of momentum, since the original scale of the fits is much closer to the lattice scale than the scale generated dynamically in the numerical calculation. Thus the rescaling procedure does not shift the momentum range to low values as much. Figures 4.35 to 4.38 show the imaginary part of the gluon fit-functions in the positive half-plane. Comparing them to the respective results obtained numerically shown in figures 4.11 and 4.12, we see that there are no significant differences, as was the case for the real part. We can therefore conclude that the gluon dressing function in the positive half-plane does not depend much on the type of solution chosen. The similarity between the fit-functions and the numerical result reassures us that the numerical result is sensible.

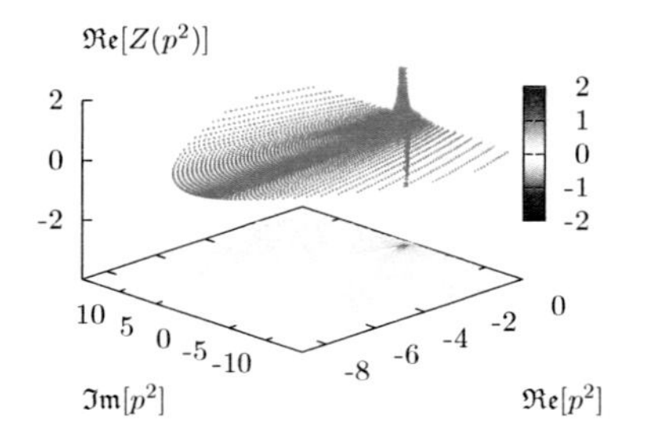

Figure 4.39: The real part of the decoupling type gluon fit-function in the negative half-plane. Perspective 1. There is a singularity at about $0.4\,\mathrm{GeV}^2$ and a cut along the negative real axis, without discontinuity.

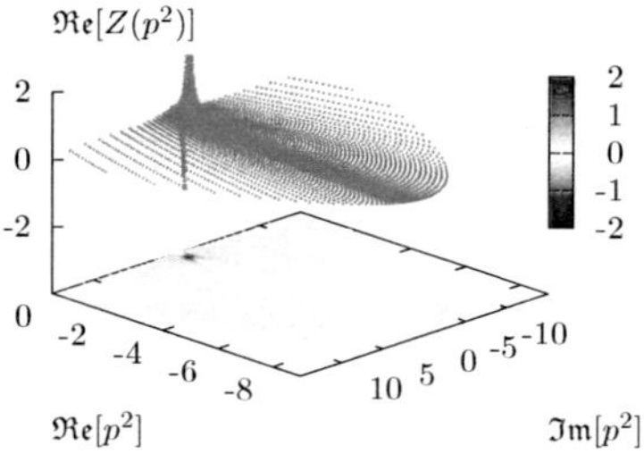

Figure 4.40: The real part of the decoupling type gluon fit-function in the positive half-plane. Perspective 2.

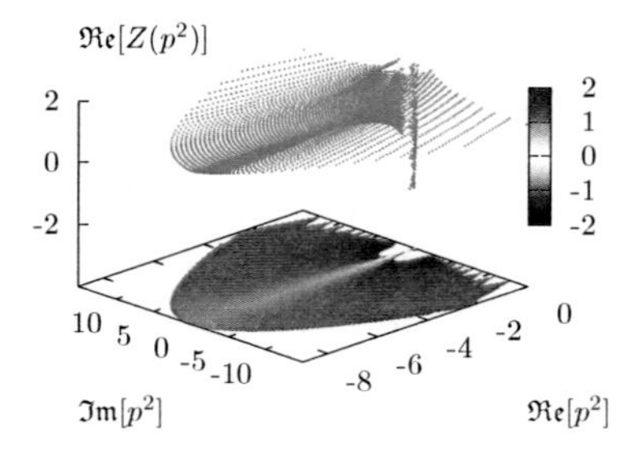

Figure 4.41: The real part of the scaling type gluon fit-function in the negative half-plane. Perspective 1. There is a singularity at about $0.3\,\mathrm{GeV}^2$ and a cut along the negative real axis without discontinuity.

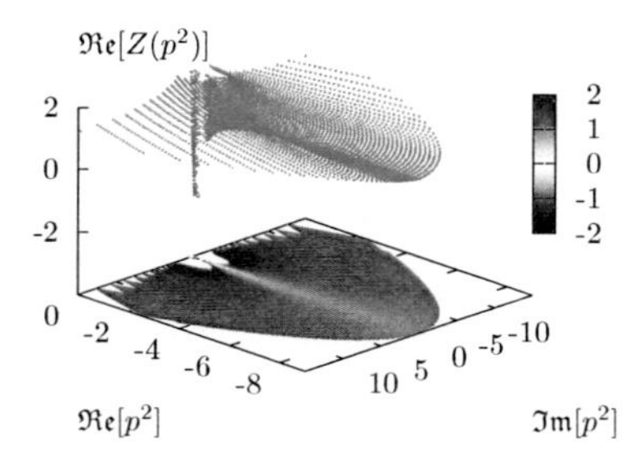

Figure 4.42: The real part of the scaling type gluon fit-function in the negative half-plane. Perspective 2.

In the negative half-plane the differences are more pronounced between the two types of fit-functions and even more so between the fits and the numerical results. Still there are similarities. It does not make sense anymore to compare the fit-functions and the numerical results regardless of the type when looking at the negative half-plane. So we will first compare the decoupling type fit-function to our numerical result and then compare the two types of fit-functions with each other.
Consider first the real part of the decoupling fit-function, which is shown in figures 4.39 and 4.40. As discussed above we find a singularity at about $0.4\,\mathrm{GeV}^2$. Apart from a shift in the position of the singularity, we find the same in our numerical result (see figures 4.13, 4.14 and 4.15. The fit-function also displays a branch-cut along the negative real axis without discontinuity, which comes from the logarithmic part of the fit. There is no cut present in our numerical result.[22] Another slight difference is, that there is some kind of dent along the negative real axis, which is not present in the numerical result. However this difference is quite small.
Comparing the fit-functions for scaling (figures 4.41 and 4.42) and decoupling, we also observe that they are rather similar. The difference is mostly a change of overall scales and a slight shift in the position of the singularity. In both cases there is a branch-cut without discontinuity and even the aforementioned dent is present in both fit-functions. This means that the numerical result is even quite similar to the scaling fit-function and we conclude that the difference between scaling and decoupling seems to be quite small for the real part of the gluon dressing function even in the negative half-plane, where significant differences could well have arisen.

[22] Our final goal is to solve the glueball BSE, where we need the solution of the Yang-Mills system in the complex plane as input. For the integrations that will be performed solving the BSE it is irrelevant, whether there is a branch-cut *without* discontinuity or not. If there is a discontinuity this is not so.

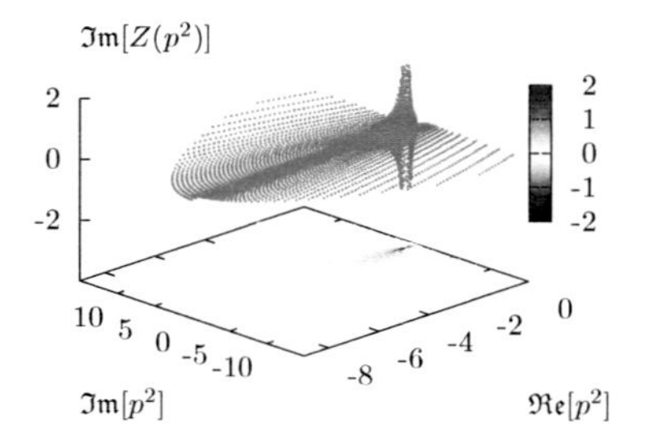

Figure 4.43: The imaginary part of the decoupling type gluon fit-function in the negative half-plane. Perspective 1. There is a singularity at about 0.4 GeV^2 and a cut along the negative real axis with a discontinuity.

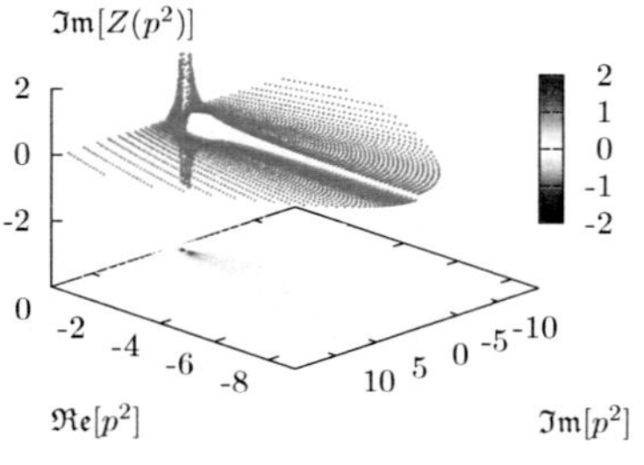

Figure 4.44: The imaginary part of the decoupling type gluon fit-function in the positive half-plane. Perspective 2.

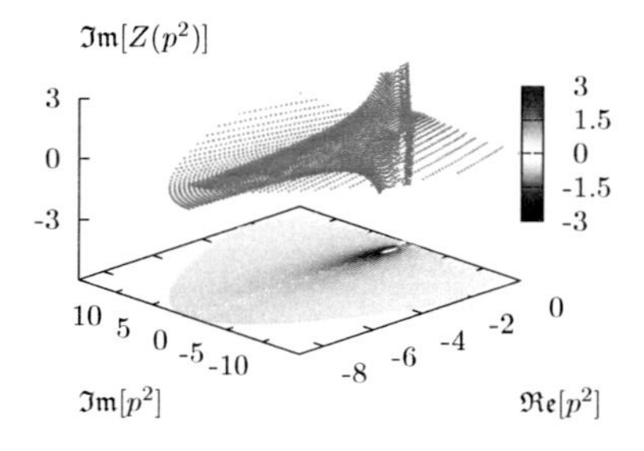

Figure 4.45: The imaginary part of the scaling type gluon fit-function in the negative half-plane. Perspective 1. There is a singularity at about 0.3 GeV^2 and a cut along the negative real axis with a discontinuity.

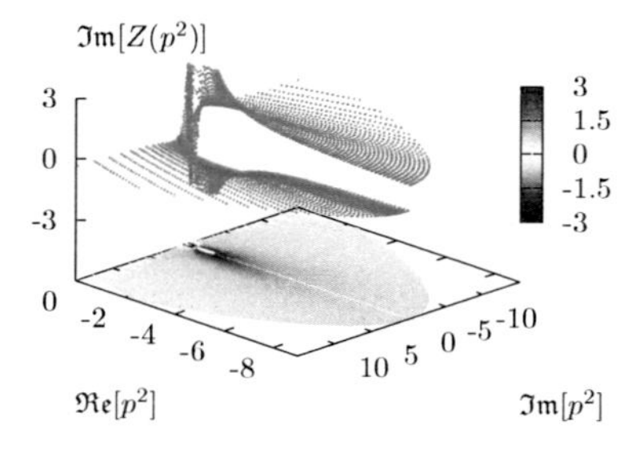

Figure 4.46: The imaginary part of the scaling type gluon fit-function in the negative half-plane. Perspective 2.

Turning to the imaginary part of the gluon fit-function, we again take a look on the decoupling type first (figures 4.43 and 4.44). We find that there is a singularity, which is odd parallel to the imaginary axis and a branch-cut with a pronounced discontinuity. In our numerical solution (figures 4.16, 4.17 and 4.18), we find a similar singularity slightly shifted in position, but no branch-cut and no discontinuity. The fact that there is no branch-cut and discontinuity might stem from numerical obstacles, but here we have to conclude, that there is a significant difference between our numerical result and the decoupling fit-function.
Looking at the scaling solution shown in figure 4.45 and 4.46 again we see, that the difference between the fit-functions for scaling and decoupling amounts to a change of scales and a shift of the position of the singularity.

The ghost fit-functions

Considering the ghost fit-function, we first stress that a comparison between fit-function and numerical result does make sense for the decoupling case only. Also the comparison between scaling and decoupling fit-function is only of limited use, since for the ghost dressing function in the infrared we have the important difference that the decoupling function is finite, while the scaling version diverges, when the momentum approaches zero. As we will see, this leads to different analytic structures in the negative half-plane.

As we did for the gluon fit-functions, we will start our discussion of the ghost fit-functions with the positive half-plane. Consider first the real part for the decoupling solution as shown in figures 4.47 and 4.48. We see that it is constant for small absolute values of the squared momentum and vanishes like a logarithm for large absolute values of the argument. Besides there is nothing noteworthy and comparing the fit-function to our numerical result as shown in figures 4.19 and 4.20, we find that both are very similar apart for some changes of overall scale. This again provides a useful cross check, for our numerical results.

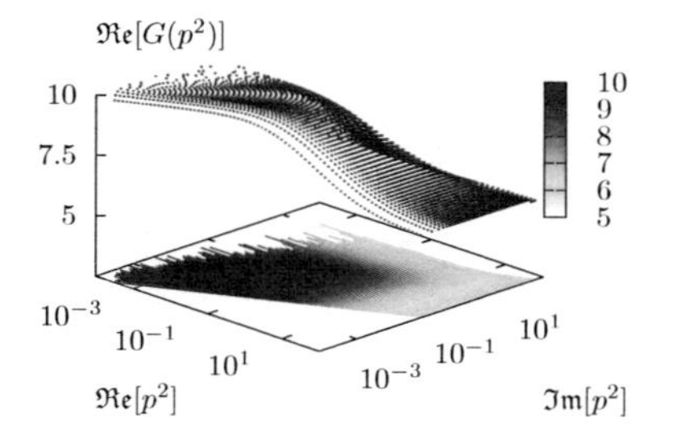

Figure 4.47: The real part of the decoupling type ghost fit-function in the positive half-plane. Perspective 1.

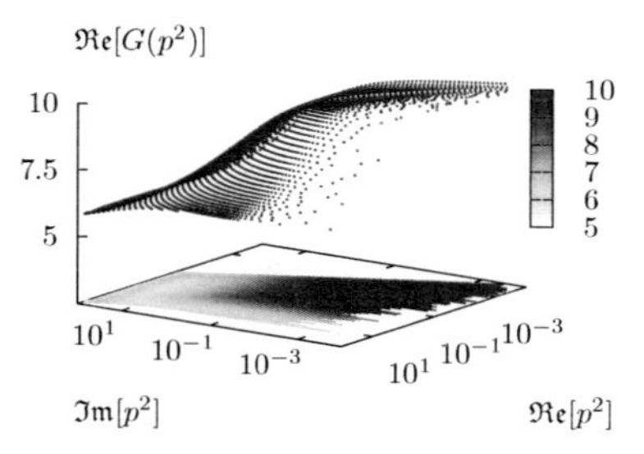

Figure 4.48: The real part of the decoupling type ghost fit-function in the positive half-plane. Perspective 2.

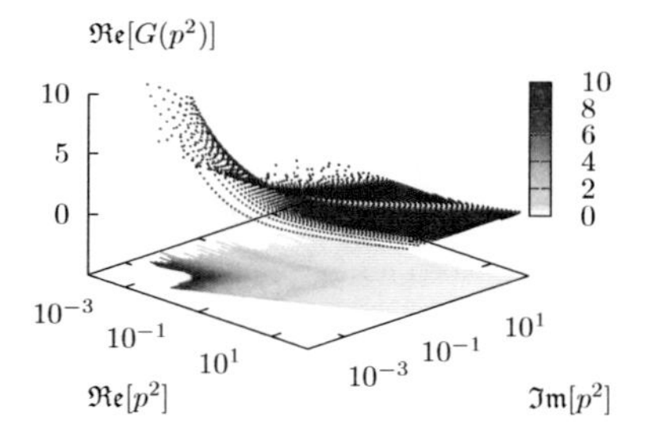

Figure 4.49: The real part of the scaling type ghost fit-function in the positive half-plane.

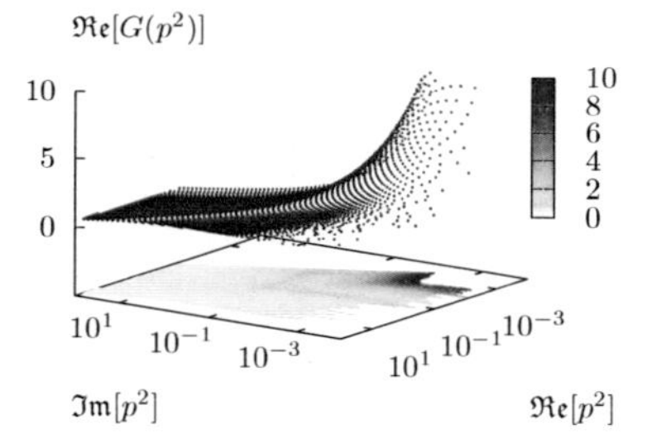

Figure 4.50: The real part of the scaling type ghost fit-function in the positive half-plane. The function value are shown on a logarithmic axis to emphasize the power-law behaviour in the infrared.

The scaling fit-function shown in figures 4.49 and 4.50 looks significantly different to the decoupling version as expected. We see the divergence in the infrared that is induced by the infrared part of the fit (cf.. eq. (4.2.19)) with $\lambda \approx 0.6$. In figure 4.50 we chose a triple-logarithmic plot to emphasize the power-law behaviour in the infrared. In the ultraviolet as expected, we see a logarithmic vanishing of the ghost fit-function.

The imaginary parts of the ghost fit-function also exhibit differences in the infrared behaviour. First consider the decoupling fit-function shown in figures 4.51 and 4.52. Actually there are no remarkable structures present. It is mostly flat and vanishing for small as well as for large momenta. Comparing it to our numerical result (figures 4.21 and 4.22), we find again that both are very similar. The ghost fit-function in the negative half-plane is more interesting. Here we will see the impact of the fundamentally different behaviour of the fit-function for small positive momenta. As before we consider first the case of decoupling. Looking at figures 4.55 and 4.56, we see that there is a finite "lump" positioned around the typical scale of decoupling at $0.4\,\mathrm{GeV}^2$. Furthermore, there is a dent along the negative real axis. Also there is a branch-cut without discontinuity along the negative real axis, which is not seen well in the figures. Comparing the fit-function again to the numerical result shown in figures 4.23, 4.24 and 4.25, we find that again both look somewhat similar. We also find a finite lump in our numerical result. However in the numerical result the lump is much steeper than in the fit-function and more importantly it is located at zero and falling rapidly along the negative real axis, which is a significant difference. We also do not see the dent in our numerical result, but that seems to be a point of minor importance.
It is hard to tell whether that is simply a numerical deviation or a feature of the solution of the Yang-Mills system. We have to recall that the ghost fit-function for decoupling is the least well justified of our fit-functions. The original fit-function was constructed from numerical data and the Schwinger-function for the scaling gluon dressing function.

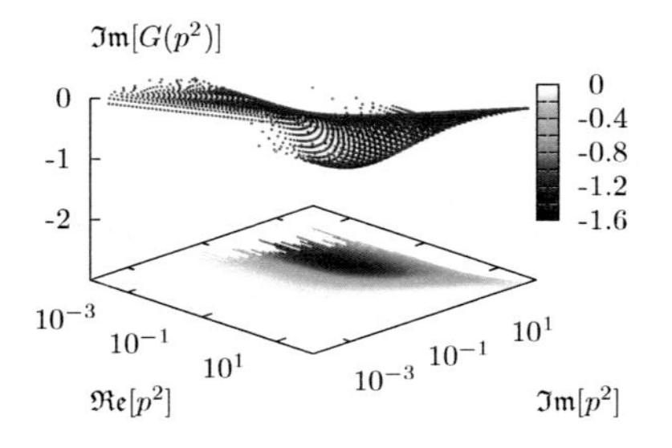

Figure 4.51: The imaginary part of the decoupling type ghost fit-function in the positive half-plane. Perspective 1.

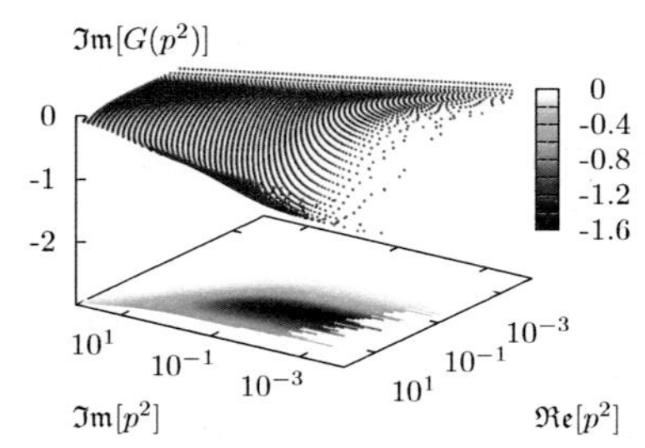

Figure 4.52: The imaginary part of the decoupling type ghost fit-function in the positive half-plane. Perspective 2.

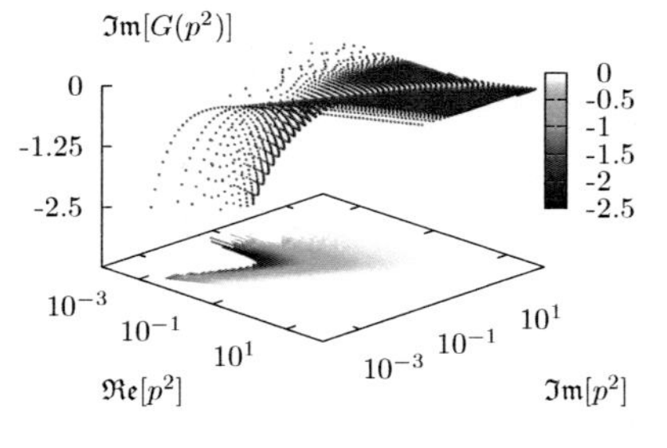

Figure 4.53: The imaginary part of the scaling type ghost fit-function in the positive half-plane. Perspective 1.

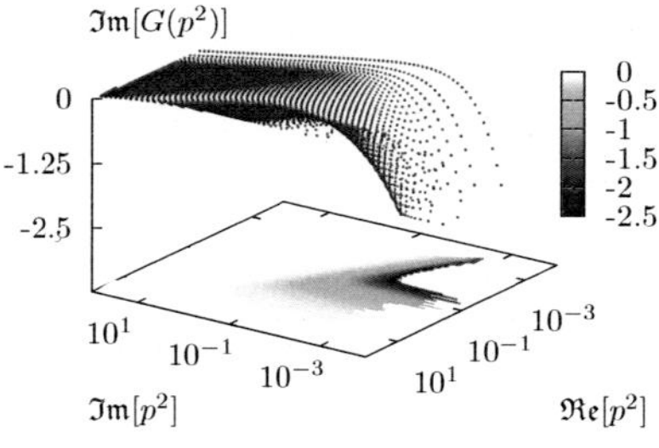

Figure 4.54: The imaginary part of the scaling type ghost fit-function in the positive half-plane. Perspective 2.

We constructed the scaling ghost function from the running coupling, but not from actual numerical data. The decoupling versions of the fit-functions furthermore was constructed by analogy, where we relied on the fact that the gluon dressing function does not change much when switching from scaling to decoupling. All in all we have to be careful with the fit-function of the ghost in the case of decoupling and thus we prefer the numerical result as the more trustworthy.
For scaling in figures 4.57 and 4.58 we clearly see the divergence at zero as expected. Note that even though on first glance the scaling result looks very similar to our numerical result, the numerical result is finite at zero. Again there is a branch-cut without discontinuity along the negative real axis, which is not visible in the plots. Also the dent as found for decoupling is present again.
The most significant differences between the fit-functions are found in the imaginary parts of the ghost fit-functions. As before consider first the decoupling case as plotted in figures 4.59 and 4.60. We find it vanishing at zero and then a finite structure, odd parallel to the imaginary axis arises, located at the typical scale for decoupling at $0.4\,\mathrm{GeV}^2$. Furthermore we have a branch-cut along the negative real axis with a discontinuity. This is quite different from our numerical result shown in figures 4.26, 4.27 and 4.28. There we find a similar finite structure, however it is located at zero momentum. It is possible that our numerical accuracy was not sufficient to make the system generate a branch-cut in the solution for the ghost dressing function. Looking at the respective plots of the numerical result for the ghost dressing function leaves the impression that something like that might be possible. Note that the numerical emergence of a branch-cut in an iterative procedure usually is a step-by-step fusion of singularities into a continuous cut. As described in section 4.2.2, in the ghost dressing functions we had lots of divergences along the negative real axis during the iterations, which diminished step-by-step. Still we have to conclude that we do not see a branch-cut with discontinuity in our numerical result and also the position of the dominant structure is different from the decoupling fit-function.

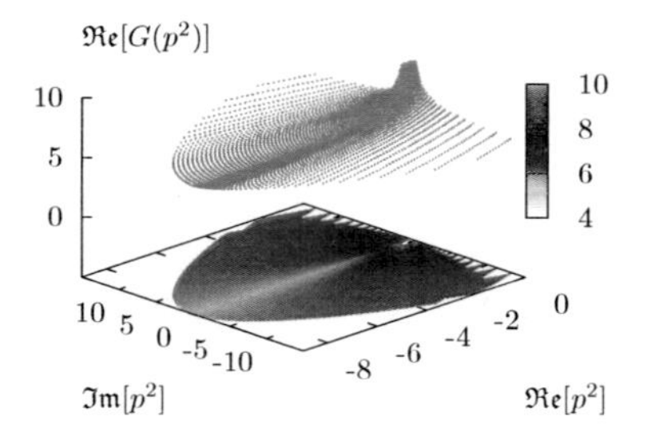

Figure 4.55: The real part of the decoupling type ghost fit-function in the negative half-plane. Perspective 1. There is a singularity at about 0.4 GeV^2 and a cut along the negative real axis without discontinuity.

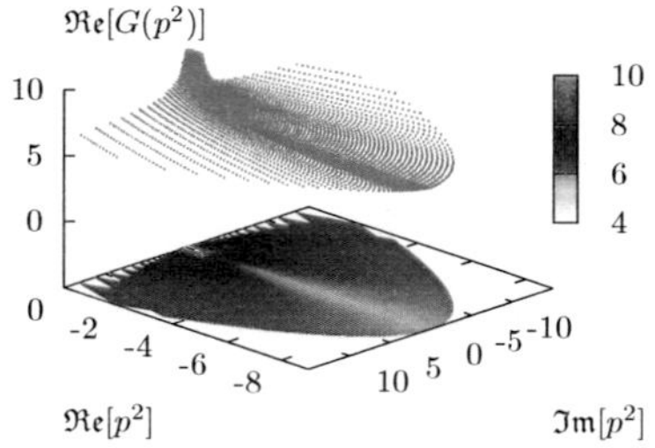

Figure 4.56: The real part of the decoupling type ghost fit-function in the negative half-plane. Perspective 2.

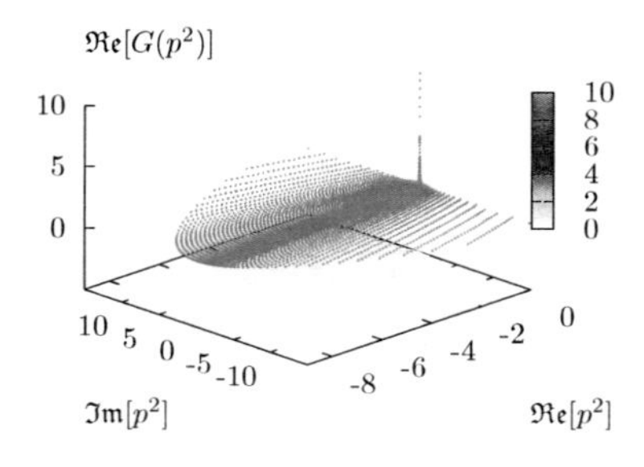

Figure 4.57: The real part of the scaling type ghost fit-function in the negative half-plane. Perspective 1. There is a singularity at the origin of the complex plane and a cut along the negative real axis without discontinuity.

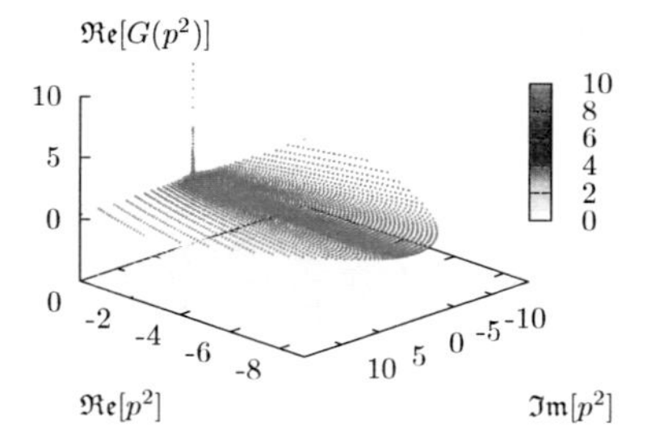

Figure 4.58: The real part of the scaling type ghost fit-function in the negative half-plane. Perspective 2.

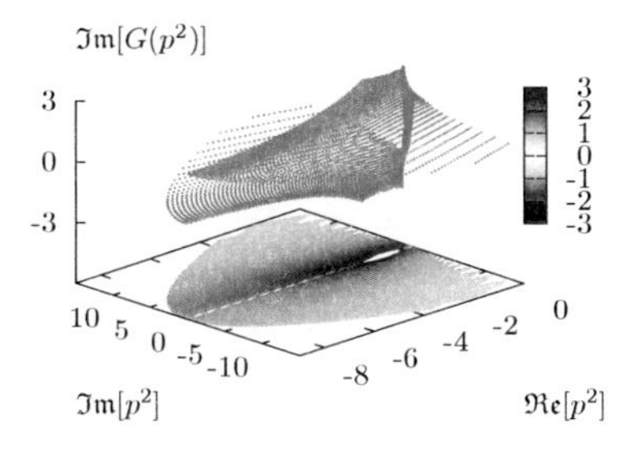

Figure 4.59: The imaginary part of the decoupling type ghost fit-function in the negative half-plane. Perspective 1. There is a singularity at about 0.4 GeV^2 and a cut along the negative real axis with a decided discontinuity.

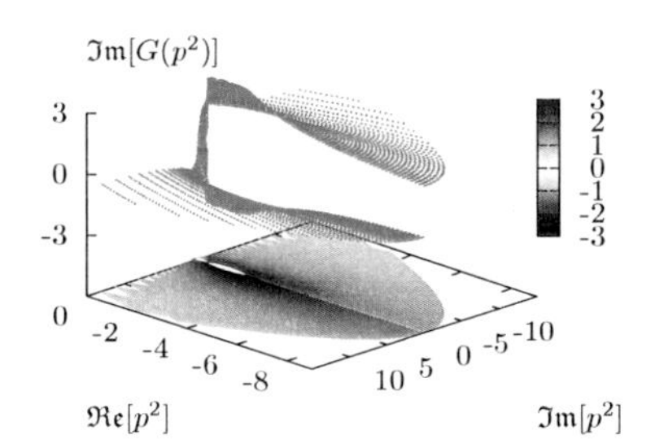

Figure 4.60: The imaginary part of the decoupling type ghost fit-function in the negative half-plane. Perspective 2.

Turning finally to the imaginary part of the scaling type fit-function for the ghost shown in figures 4.61 and 4.62, we find divergence at zero. This divergence is odd parallel to the imaginary axis. So again the position of the divergence is like in our numerical result, however being a decoupling type solution our numerical ghost dressing function is finite. Furthermore as expected there is a branch-cut along the negative real axis with a discontinuity.

In this chapter we have discussed in detail the solution of the Yang-Mills system with special emphasis on the solution for complex momenta. We have also discussed fit-functions for ghosts and gluon. The importance for our purposes of the dressing functions of ghosts and gluons is due to the fact that we need them as input in the BSEs describing glueballs. However the dressing functions of ghosts and gluons are interesting on their own, since they encode important physical information for instance about confinement. We have applied the shell-method to determine the solution of

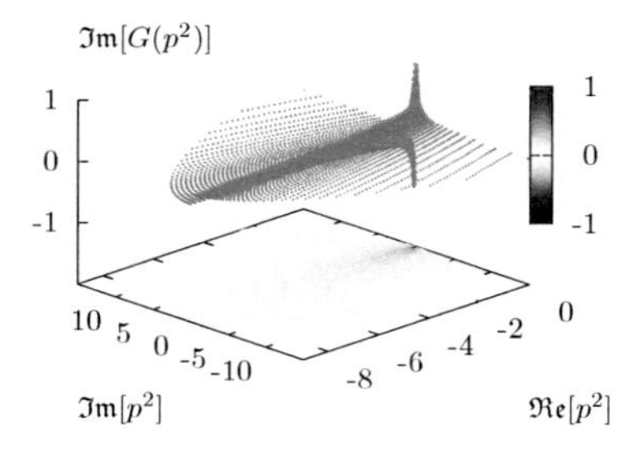

Figure 4.61: The imaginary part of the scaling type ghost fit-function in the negative half-plane. Perspective 1. There is a singularity at the origin of the complex plane and a cut along the negative real axis with a slight discontinuity.

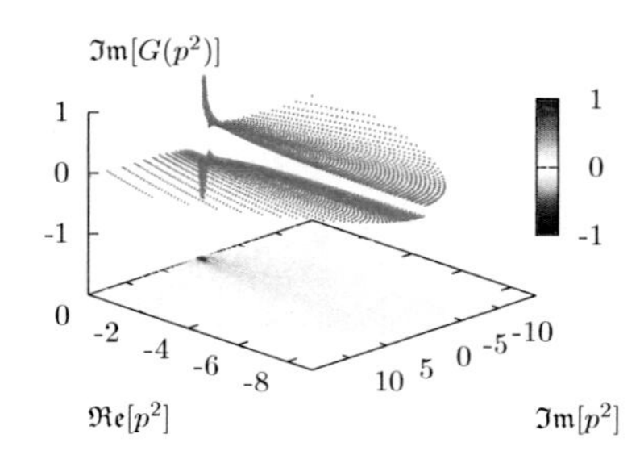

Figure 4.62: The imaginary part of the scaling type ghost fit-function in the negative half-plane. Perspective 2.

the Yang-Mills system in the complex plane numerically. We had to employ a certain truncation scheme to make the Yang-Mills system numerically tractable. However the truncation scheme was not only chosen such as to render the equations solvable, but also to preserve all known symmetries. Yet whether or not the truncation is justified and useful, is decided in the application of the scheme to the BSE, where we also have to employ a consistent truncation. Since the glueball BSE provides the connection from our ghosts and gluons to observable quantities, any truncation scheme that is to produce "physical" results has to be chosen such as to produce a reasonable mass spectrum from the BSE. The truncation scheme we employed here, is only a first step. It is the simplest that preserves all symmetries and leads to sensible ghost and gluon dressing function. The dressing functions obtained numerically are rather similar to the fit-functions we discussed in the second part of our treatment of the ghost and gluon dressing functions in the complex plane. This confirms our choice of truncation as being sensible also, if

only a first try. Anyhow the Yang-Mills system and its solutions in the complex plane, will remain an interesting and viable topic in continuum QCD. We have presented the main lines on how to approach the system for complex momenta.
In the following chapter we will finally turn to the numerical investigation of the glueball BSE. We now have all ingredients at hand, the necessary equations, the amplitudes corresponding to a given quantum number of the glueball and the non-perturbative dressing functions for the constituents (i.e. ghosts and gluons).
First we will discuss general properties of glueballs. Then we will present the concrete BSE framework we choose and the truncation. We will describe in detail how the solution of the glueball BSE is done numerically. Finally we will present the results for various glueball states and discuss the quality of our truncation scheme.

Chapter 5

Numerics of the glueball Bethe-Salpeter equations

5.1 Glueballs

In this section we will give an overview about glueballs and what has already been done to study their properties. Physical states comprised of gluons have first been studied in the middle of the 1970s. The first paper with a more detailed account of properties of gluonic bound states (or *gluonium*) was [FM75]. In the early studies the starting points were the MIT Bag Model [CJJ$^+$74, JJ76] and the OZI rule [Zwe64, Oku63, IOS66] where the possible pole-structure in gluonic decay channels of narrow hadron resonances was speculated to be a new kind of meson[1] connected to the pomeron and possibly of purely gluonic nature [FN75, BGPP75, BPP76, CFM80].
The first study of gluonic bound states on a lattice presumably was [KSS76], where such states were called *boxcitons*,[2] because it is generated by evaluating expectation values along closed and connected sets of link variables, which in three-dimensions in the simplest case form something like a box. The name glueball at first summarised all possible purely gluonic bound states. Today this term is mostly used for states comprised of two

[1] The meson was called *O-meson* by the authors.

[2] In later works the name *boxiton* can also be found.

gluons. It is commonly suspected that glueballs are connected to the pomeron which is found in Regge theory. The pomeron was introduced by Gribov in 1961, when he found that the, at that time commonly accepted, picture of diffractive scattering in particle collisions at high energy was inconsistent with unitarity and analyticity conditions of the scattering amplitudes [Gri61]. One important consequence is that the cross-sections of proton-proton- and proton-antiproton scattering should be equal at high energies. This had already been predicted on analyticity considerations by Pomeranchuk [Pom58], which is the reason for the name pomeron. The pomeron has positive charge-conjugation parity, which is also true for glueballs. In analogy to the *odderons*, which are the negative charge-conjugation parity counterparts of pomerons, states made of three gluons are sometimes called *oddballs*,[3] since these states also have negative charge-conjugation parity. States with more than three gluons often are called *gluelumps*.
In the last about fifteen years there has been done quite some work to study properties of glueballs. While the early calculations mostly found the low-lying glueball states to have masses below 1.5 GeV, today it is commonly agreed that the lowest states have larger masses.
The most reliable approaches to glueball properties possibly are lattice gauge theory [CAD$^+$06, MT05, RIGM10, McN09, HMMP06, B$^+$93, LWC02, LW02] and (effective) Hamiltonian calculations in Coulomb gauge [SS03, LEBC06, RSS$^+$99, LECdAB$^+$02, GSG$^+$08, KS00], some of which also make use of Regge theory. There is a variety of other methods like sum-rules, supersymmetric theories, gauge-gravity duality etc. however they are suffering from the fact that always some assumptions have to be made out of hand. In case of sum-rules, assumptions about the glueball spectrum have to be made a priori, while for gauge-gravity duality or supersymmetric theories, prescriptions how to transfer results into physical relevance have to be assumed. Therefore we will not refer to such methods in the following and concentrate on methods which are based on rather solid physical ground.

[3] Presumably first mentioned in [CCF$^+$81]. However it is unclear whether the authors removed the notion of oddballs in proof.

Lately some work has been done on semirelativistic constituent gluon models [Bui07, MSSB08], where a set of parameters in the Hamiltonian is fitted to lattice results. Also there has been presented a completely new approach using improved Gribov-Zwanziger theory [DGS11].
We will not go into detail of each approach. For a recent review see [MKV09]. Rather we will collect the results found in the literature mentioned above and make some comments.

Let us first shortly comment on results from lattice gauge theory. Calculating glueball mass spectra in principle is an easy task on the lattice. In the simplest approach it is not much more than calculating expectation values over correlators of link variables and sorting out symmetry properties to determine which quantum number the expectation value corresponds to.
In principle such a calculation is done without fixing the gauge. There is no need to fix the gauge, calculating expectation values connected to observables which are gauge-independent. Thus carried out lattice calculations are insensitive to subtleties connected to gauge-fixing, like whether the dressing function correspond to scaling or decoupling solutions.[4] However in lattice calculations it is not possible to determine "how many" gluons comprise a bound-state under investigation.[5]
Looking at tables 5.2, 5.4 and 5.3, we see that the states having spin one are heavier than one would naively expect. The reason is that there are no bound states of two gluons[6] that have total spin one. This is known as *Yang's theorem* or *Landau-Yang theorem* and has been proven in [Lan48] and [Yan50]. To see that, it is most convenient to use the helicity formalism of Jacob and Wick [JW59]. Amplitudes of two massless vector particles (photons, gluons etc.) have to satisfy parity-conservation, which in the helicity

[4] Which on the lattice would correspond to the question whether the gauge-fixing procedure leads to ensembles representing a continuum gauge-fixing of scaling or decoupling type. Gauge-fixing on the lattice is a highly non-trivial matter and a field of active research.

[5] The number of gluons involved is not a well-defined notion. It is a general feature that the number of strongly interacting particles contained in a bound state is scale-dependent if sensible at all. The same holds e.g. for mesons and baryons. Perturbatively they contain a rather well-defined number of constituents, non-perturbatively this notion becomes unclear. In a sense one might say that this is even more so for gluons.

[6] Or more general: of two massless vector particles

formalism can be written as[7]

$$F^J_{\lambda_1,\lambda_2} = \eta_P \, (-1)^J \, F^J_{-\lambda_1,-\lambda_2}, \tag{5.1.1}$$

where we have the two-vector-particle amplitude $F^J_{\lambda_1,\lambda_2}$, the helicities of the two vector particles λ_1 and λ_2 and the intrinsic parity of the bound state η_P. Furthermore such amplitudes have to be Bose-symmetric, which entails

$$F^J_{\lambda_1,\lambda_2} = (-1)^J \, F^J_{\lambda_2,\lambda_1}. \tag{5.1.2}$$

To understand the latter identity, consider a system of two massless vector particles in their center-of-mass frame. The system can be described by a tensor $T_{ij}(x)$, where x is the relative momentum of one of the vector particles. If we exchange the vector particles, the relative momentum picks up a minus sign. Since the angular momentum tensors, which are included in the tensor T_{ij} depend on J relative momenta, we immediately find relation (5.1.2).
From eqs. (5.1.1) and (5.1.2), we find for two particles of the same helicity[8]

$$F^J_{\lambda,\lambda} = (-1)^J \, F^J_{\lambda,\lambda} \tag{5.1.3}$$

and

$$F^J_{\lambda,-\lambda} = \eta_P \, (-1)^J \, F^J_{-\lambda,\lambda} = (-1)^J \, F^J_{-\lambda,\lambda}, \tag{5.1.4}$$

where we find the last equality using (5.1.2). From these equations we can easily find what is called "natural states" in [FM75]. We see that, if J is odd, $F^J_{\lambda,\lambda} = 0$ and if $\eta_P = -1$, $F^J_{-\lambda,\lambda} = 0$. Thus J^{-+} states vanish for odd J. Also J^{++} states vanish if J is odd and $\lambda_1 = \lambda_2$, because $F^J_{\lambda,\lambda} = 0$ for odd J. Furthermore we note that there are no amplitudes $F^J_{\lambda,\lambda}$ for $J = 0$ and $J = 1$, for such tensors represent at least a system of total spin $J = 2$.

[7] This derivation along with a detailed treatment of quarkonia coupling to two photons can be found in [Chu02].

[8] Massless particles always have exactly two helicity polarisations. Thus there are two states comprised of two massless particles with the same helicity: $\lambda_1 = \lambda = \lambda_2$, $\lambda_1 = \lambda = -\lambda_2$. Usually these two states are denoted with the sum of their helicity polarisations $\Lambda = \lambda_1 + \lambda_2$, so in the first case we have $\Lambda = 2$ and in the second case $\Lambda = 0$.

Since states of two massless vector particle necessarily have positive charge-conjugation parity [Lan48], we see that there are no $J = 1$ states left. Coming back to consider only gluons, we find the natural two-gluon bound state quantum numbers in table 5.1.

0^{++}, 0^{-+}
2^{++}, 2^{++}, 2^{-+}
3^{++}
$\dots$
$(2n)^{++}$, $(2n)^{++}$, $(2n)^{-+}$
$(2n+1)^{++}$.

Table 5.1: The natural quantum numbers of two-gluon bound states.

There are states that occur twice in table 5.1, if $J > 1$ and even. This due to the existence of a fundamental spin two combination of to gluons. We also found a corresponding tensor (see eq. (3.3.12)). From that tensor we will get an additional structure in the corresponding Bethe-Salpeter amplitude (see eq. (3.3.47)). In our formalism however, we get additional contributions to states with odd J also, which are absent in table 5.1. Due to the fact that both representations (ours and the direct products of the representation of the constituent particles) are inequivalent, that however is not a problem. However we have already mentioned in section 3.3.1 that in our explicitly covariant representation of the Bethe-Salpeter amplitudes, some properties, found in non- or semirelativistic spin (or helicity) treatments, are not obvious or even absent.[9] We can construct $J = 1$ amplitudes easily. Also we can construct 3^{-+}, 5^{-+}, $\dots$ states. We will consider some such states but keep in mind that they might not really exist. Especially for $J = 1$ we will find a state much lighter than is found e.g. on the lattice.

[9] This is no surprise. It is also no problem to construct explicitly covariant amplitudes for mesons with quantum numbers considered exotic from the view-point of the constituent quark model.

	masses (GeV)					
J^{PC}	lattice		Hamiltonian/ Regge theory		constituent models	
0^{++}	1.71	[CAD+06]	1.98	[SS03]	1.71	[Bui07]
	1.55	[B+93]	1.58	[KS00]	1.60	[MSSB08]
	1.75	[LWC02]			1.86	[MSSB08]
	1.73	[MP99]				
$0^{++\prime}$	2.67	[MP99]	3.26	[SS03]	2.63	[Bui07]
			2.71	[KS00]	2.59	[MSSB08]
					2.99	[MSSB08]
0^{-+}	2.56	[CAD+06]	2.22	[SS03]	2.61	[Bui07]
	2.33	[B+93]	2.56	[KS00]	2.17	[MSSB08]
	2.59	[MP99]			2.49	[MSSB08]
$0^{-+\prime}$	3.64	[MP99]	3.42	[SS03]	3.53	[Bui07]
			4.15	[KS00]	3.23	[MSSB08]
					3.71	[MSSB08]
0^{+-}	4.78	[CAD+06]			4.04	[MSSB08]
	4.74	[MP99]			4.66	[MSSB08]

Table 5.2: Glueball masses for total spin $J = 0$ from various different studies. We did not include the error bars, for they are of minor interest here and not always included in the original work.

The second anomaly in the glueball spectrum is that the 0^{-+} state is heavier than the 2^{++} state. From meson spectra, which are physically the closest relatives to glueballs, nothing such is known. One would expect that a state having spin two should be heavier than a pseudoscalar state. This anomaly has not been puzzled out so far. During our work we obtained evidence that the reason for this anomaly might be connected to contributions from ghost states in Landau gauge (cf. section 5.3). On the lattice such a mechanism is invisible, since one does a summation over ensembles of fields without gauge-fixing and thus mechanisms, connected to gauge-fixing and leading to the anomalous mass hierarchy under consideration are hidden.

	masses (GeV)					
J^{PC}	lattice		Hamiltonian/ Regge theory		constituent models	
2^{++}	2.39	[CAD+06]	2.42	[SS03]	2.53	[Bui07]
	2.23	[B+93]	2.59	[KS00]	2.05	[MSSB08]
	2.31	[B+93]			2.38	[MSSB08]
	2.42	[LWC02]				
	2.40	[MP99]				
$2^{++\prime}$			3.11	[SS03]	2.97	[Bui07]
			3.73	[KS00]		
2^{-+}	3.04	[CAD+06]	3.09	[SS03]	2.97	[Bui07]
	2.95	[B+93]			2.57	[MSSB08]
	3.08	[B+93]			2.98	[MSSB08]
	3.10	[MP99]				
$2^{-+\prime}$	3.89	[MP99]	4.13	[SS03]	3.76	[Bui07]
					3.35	[MSSB08]
					3.86	[MSSB08]
2^{+-}	4.23	[CAD+06]			3.91	[MSSB08]
	3.39	[B+93]			4.54	[MSSB08]
	4.40	[B+93]				
	4.14	[MP99]				
2^{--}	4.01	[CAD+06]	3.71	[KS00]	4.42	[MSSB08]
	3.83	[B+93]			5.13	[MSSB08]
	4.05	[B+93]				
	3.93	[MP99]				

Table 5.3: Glueball masses for total spin $J = 2$ from various different studies. We did not include the error bars, for they are of minor interest here and not always included in the original work.

	masses (GeV)					
J^{PC}	lattice		Hamiltonian/ Regge theory		constituent models	
1^{++}	3.96	[B+93]				
1^{+-}	2.98	[CAD+06]			3.99	[MSSB08]
	2.94	[MP99]			4.63	[MSSB08]
	2.90	[B+93]				
1^{--}	3.83	[CAD+06]	3.49	[KS00]	3.43	[MSSB08]
	4.36	[B+93]			4.00	[MSSB08]
	3.85	[MP99]				

Table 5.4: Glueball masses for total spin $J = 1$ from various different studies. We did not include the error bars, for they are of minor interest here and not always included in the original work.

In Coulomb gauge studies there might well be a different mechanism, however the investigations of the glueball mass spectrum conducted so far did not explicitly consider ghost and gluon contributions separately. Also using Regge theory or effective models will hide such mechanisms. In this respect our approach turns out to be advantageous. Since we deal with the fundamental degrees of freedom of QCD, we are able to obtain information about the concrete mechanisms leading to the observed mass spectrum of glueballs. Of course so far we did only find hints, that this anomaly might be connected to ghosts. It would be worthwhile to study the effect of ghosts in more detail, using more sophisticated truncation schemes. Also it would be very interesting to investigate, how the anomalous mass hierarchy comes about in the Gribov-Zwanziger framework, where gauge fixing is a prominent ingredient of the theory. In total from table 5.2 to 5.6 it can be seen that for the mass spectrum of glueballs there are quite solid results. Especially looking at the findings of the more recent lattice calculations, we see that the results of different studies agree very well. Taking also Hamiltonian or Regge theory calculations into account a rather clear spectrum arises.
The constituent models also agree quite nicely, one should however note that these are fitted to lattice data. More interesting insight can be drawn from constituent model

	masses (GeV)					
J^{PC}	lattice		Hamiltonian/ Regge theory		constituent models	
3^{++}	3.67	[CAD+06]	3.33	[SS03]	3.13	[Bui07]
	3.92	[B+93]	3.58	[KS00]	3.13	[MSSB08]
	3.69	[MP99]			3.61	[MSSB08]
$3^{++\prime}$			4.29	[SS03]	3.89	[Bui07]
3^{+-}	3.60	[CAD+06]			4.03	[MSSB08]
	3.55	[MP99]			4.57	[MSSB08]
3^{--}	4.20	[CAD+06]	4.03	[KS00]	3.57	[MSSB08]
	4.13	[MP99]			4.17	[MSSB08]

Table 5.5: Glueball masses for total spin $J = 3$ from various different studies. We did not include the error bars, for they are of minor interest here and not always included in the original work.

calculations for other glueball properties e.g. decay widths, which are accessible using such models. Furthermore such models allow for investigations of glueballs at finite density and temperature, which make it possible to study their effect for instance in quark-gluon plasmas[BB09, Bui10].

All together we can conclude that the mass spectrum of glueballs is a excellent benchmark to test our formalism. Since we use the probably most simple truncation scheme, we do not expect that we will reproduce the given spectrum in much detail. To achieve that, more work on improving the truncation is necessary, which we however leave for future investigations. We want to stress the advantages of the approach presented in this work again. Firstly being an approach using the fundamental degrees of freedom of QCD, we are able to obtain information about physical mechanism, which lead to the concrete structure of the mass spectrum of glueballs and in future applications to an insight to the formation of gluonic bound states, in the gauge chosen, which in our case is Landau gauge.

	masses (GeV)					
J^{PC}	lattice		Hamiltonian/ Regge theory		constituent models	
4^{++}	3.65	[LW02]	3.99	[SS03]	3.60	[Bui07]
$4^{++\prime}$			4.28	[SS03]	3.78	[Bui07]
4^{-+}			4.27	[SS03]	3.78	[Bui07]
$4^{-+\prime}$			4.98	[SS03]		

Table 5.6: Glueball masses for total spin $J = 4$ from various different studies. We did not include the error bars, for they are of minor interest here and not always included in the original work.

Secondly our method is readily generalised to include quark contributions and effects coming from meson/glueball-mixing. Thirdly the numerical calculation is only moderately cpu-intensive, which is of even greater importance with respect to the second advantage of quarks being easily included in the formalism. Though the calculation of glueball masses is merely a warm-up exercise on the lattice, including quarks is a formidable problem. Quarks are notoriously heavy in lattice calculations. There are lattice-actions, which allow for light quarks, but they are extremely costly in terms a cpu-time. Therefore a lattice calculation including light quarks has not been done yet. There are unquenched calculations using staggered quarks, but they are still not really light (corresponding pion masses are about 300 MeV) [RIGM10]. In our approach unquenching effects and glueball/meson-mixing are included quite naturally looking at equations (3.2.37), (3.2.38) and (3.2.37). Especially the computational power necessary for such a calculation is much smaller than for a comparable lattice calculation.
Including quarks should in principle be possible in Hamiltonian calculations as conducted in [SS03], where a variational method was applied to the gluon wave-functions in order to find eigenstates of the Hamiltonian and to determine their properties. To extend the Hamiltonian to include quarks and to do such a variational procedure with the corresponding Hamiltonian again, appears possible. However the calculations will become rather lengthy and complicated.

We will now proceed to present the assembly of our formalism into closed form. We will formulate the Bethe-Salpeter equations for glueballs and discuss the numerical treatment in some detail in the next section. Following that, in section 5.3 we will present numerical results and compare them to the literature.

5.2 The Bethe-Salpeter equation for glueballs

In this chapter we will finally apply the formalism developed in the previous chapters to tackle the glueball BSE. With the framework developed in chapter 3 and the numerical results for the dressing functions of ghosts and gluons for complex momenta, that we calculated in chapter 4, we will be able to do an attempt on solving the coupled system of BSEs for bound states of ghosts and gluons. First we will give details about how a BSE can be brought into a numerically feasible form and how it is solved. We will discuss basic techniques as well as some numerical subtleties.
The combined SDE/BSE approach has been used successfully in meson studies several times. The method almost exclusively applied in these studies was to Wick-rotate the total momentum in the BSE and to continue the DSE into the complex momentum plane.
An important property of the BSE is that it is invariant under Poincaré transformations. It is thus possible to pick a convenient frame and perform the calculation there. However numerically one usually has to resort to some kind of regularisation scheme, that can possibly break Poincaré invariance. For instance sharp cutoffs, which will be used in the numerical investigations in this work, do break Poincaré invariance explicitly. However after suitable renormalisation, the cutoffs are at least in principle removed, so that the result will be valid and invariant again.
Looking more closely on the BSE and inspecting the concrete flow of momenta through the loop, one finds that there is an undetermined parameter, because the fraction of total momentum that is carried by one constituent particle and denoted η in the following is not fixed *a priori*. To clarify matters let us take a small detour to meson BSE studies.

We show the meson BSE again, with explicit momentum routing

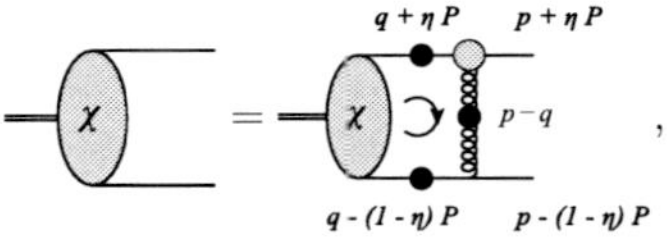

$q+\eta P$ $\quad p+\eta P$

$\chi = \chi \quad p-q \quad , \qquad (5.2.1)$

$q-(1-\eta)P$ $\quad p-(1-\eta)P$

where P is the total momentum vector, p the relative momentum vector and q is the loop momentum.

In a calculation, where Poincaré invariance is not broken at any stage, fixing that partition cannot have any impact on the result. If however Poincaré invariance is broken by the regularisation procedure, fixing the momentum partition parameter η will introduce another source of breaking Poincaré invariance which is *not* removed by renormalisation. The question how large the impact of the choice of η on the result is can be checked by simply varying it. For several mesons this has been done e.g. in [AWW02] and it turned out that in some region around $\eta = 0.5$ the impact is rather small.

The choice $\eta = 0.5$ seems natural if the masses of the constituents are equal. The constituent masses themselves are another source of problems, connected with regularisation. Looking at (5.2.1) and (2.3.19) we see that the constituent quarks will be described by functions $A(q+\eta P)$, $B(q+\eta P)$ and $A(q-(1-\eta)P)$ and $B(q-(1-\eta)P)$ respectively. In the case of equal constituent masses it turns out that this symmetry leads to a BSE which is rather insensitive to η around $\eta = 0.5$. This changes however if the masses are different.

We will see in the following that the integration probes the complex momentum plane, because we will Wick-rotate the bound state momentum-vector P. The integration will take place on parabolae and the parameter η has an impact on the position of these parabolae. Thus the insensitivity of the BSE to η in the symmetric case stems from the fact, that the position of the parabolae does not influence the integral of the BSE much, if it is close enough to $\eta = 0.5$.

However for unequal masses the different dressing functions for the quarks induce an asymmetry themselves, which can be amplified by the momentum partition. On the other hand the momentum partition parameter can be fixed such as to minimise the asymmetry coming from the unequal quarks. In fact for unequal quark masses simply taking $\eta = 0.5$ will not reproduce experimental data accurately. If one instead fixes the momentum partition for a given combination of quark masses for one particular quantum number, the spectrum of bound state masses, will be well reproduced for other quantum numbers.

From this consideration we see that the choice of the momentum partition parameter η can have quite some impact on the BSE for different constituent types. In the following we will study a system where the constituents are of the same type, since glueballs are bound states of two gluons. We thus will adopt the symmetric choice of momentum partition $\eta = 0.5$.

As is done in the meson BSE studies, we will employ a Wick-rotation. Instead of applying it to the loop momentum as is commonly done in perturbation theory, we will do a rotation of the total momentum vector if we are probing the negative $P^2 = M^2$ region.

$$(M, 0, 0, 0) \rightarrow (iM, 0, 0, 0). \tag{5.2.2}$$

This choice leads to a much simpler calculation as the rotation of the loop and relative momenta. Furthermore it allows to easily scan the solutions of the BSE depending on P^2 contiguously from the positive to the negative euclidean momentum region, if necessary. With these choices, we find that for negative euclidean bound state momenta, the dressing functions of the constituents depend on complex squared momenta

$$\begin{aligned}
p_+^2 &= p^2 + i\,|p||P|\,\cos(\theta_p) - |P|^2/4 \\
p_-^2 &= p^2 - i\,|p||P|\,\cos(\theta_p) - |P|^2/4 \\
q_+^2 &= q^2 + i\,|q||P|\,\cos(\theta_q) - |P|^2/4 \\
q_-^2 &= q^2 + i\,|q||P|\,\cos(\theta_q) - |P|^2/4
\end{aligned} \tag{5.2.3}$$

where θ_x is the angle between the total momentum and a vector x. For fixed total momentum P these momenta come to lie on parabolae in the complex momentum plane, with vertices given by $-|P|^2/4$. Furthermore it turns out that for any $0 \leq \theta_x \leq \pi$ all momenta belonging to a fixed P^2 and variable relative or loop momentum lie on parabolae, within the corresponding parabola with $\theta_x = 0$. For the numerical investigation this is very important, since it allows for a powerful way to continue dressing function of constituents into the complex plane, which is necessary for the calculation of the BSE. This method is described in detail in section 4.2.

The BSE is a homogeneous integral equation of Fredholm type. There is a lot of theory available for this kind of equations (e.g. [Tri85, Hac95]), however we are interested in the concrete numerical solutions of the BSE. The two most important entities, which can be derived from the BSE are the positions of bound state poles on the negative real total momentum axis and the Bethe-Salpeter wave functions (3.2.4). Being a first study of the BSE of glueballs we will confine ourselves to investigate formulation and justification of the corresponding equations and to approximately determine the position of bound state poles. To locate these poles there are several methods. The basic idea is the following. Consider a slightly modified version of the BSEs

$$\chi = \lambda \cdot \chi \,, \tag{5.2.4}$$

Here we have introduced a free parameter λ. This equation is a functional eigenvalue equation, which is equal to the BSE for $\lambda = 1$. We can consider both sides of (5.2.4) as dependent on P^2. In an obvious operator notation we then have

$$\Gamma(P^2, p^2, P \cdot p) = \lambda(P^2)\, K(P^2, p^2, q^2, P \cdot p, P \cdot q)) \cdot \Gamma(P^2, q^2, P \cdot q). \tag{5.2.5}$$

This equation will become the BSE at any bound state mass M_B^2, thus $\lambda(P^2 = -M_B^2) = 1$. We can use this to determine the bound state masses. We calculate equation (5.2.5) for different total momenta, determine the corresponding eigenvalues $\lambda(P^2)$ and look for the momenta at which the eigenvalue becomes one. Numerically this is done by a suitable discretisation of the integral operator $K(P^2, p^2, q^2, P \cdot p, P \cdot q))$ in (5.2.5). The discretisation with respect to the momenta is rather obvious. One discretises the external momentum p and the loop momentum q on the same grid, matching discretised momenta. The grid can be a standard Gauss-Legendre quadrature formula. To increase precision it is in principle possible to use a finer grid for the loop momenta. Therefore one uses an interpolation such as cubic splines on the external momentum points p_i and employs a finer Gauss-Legendre quadrature for the loop momentum points q_k, which are determined from the splines on p. For the angular integrals there are two major possibilities, how to discretise the angular points.

The possibly most obvious choice is to also apply a Gauss-Legendre quadrature on the angular integral, leading a totally matched momentum/angular-grid in external momentum p and loop momentum q. It is possible to use an interpolation strategy for the angular discretisation also, thus interpolating in a set of discretised angles between external momentum p and total momentum P to obtain a larger set of quadrature points in the angle between the loop-momentum q and the total momentum P^2. The alternative is to employ a Chebyshev approximation to the angular integral. The Chebyshev polynomials of the first kind[10] are defined by [GR80]

$$T_n(\cos(x)) = \cos(nx). \tag{5.2.6}$$

The zeros of the polynomial are found to be

$$\xi_{k,n} = \cos\left(\frac{(k-1/2)\pi}{n}\right). \tag{5.2.7}$$

A given function $f(x)$ can be approximated with Chebyshev polynomials by

$$f(x) \approx \sum_{k=0}^{n-1} c_k\, T_k(x) - \frac{1}{2}\, c_0, \tag{5.2.8}$$

where the coefficients are calculated as

$$c_j = \frac{2}{n} \sum_{k=1}^{n} f(\xi_{k,n})\, T_j(\xi_{k,n}). \tag{5.2.9}$$

The BSE is then discretised in the coefficients c_j on both the external and the loop momenta. The angular part of the loop integral can be calculated leaving the set

[10] One might wonder, why we do not choose the Chebyshev polynomials of the second kind and directly use them for the quadrature rule. This indeed leads formally to a slight simplification since such a Gauss-Chebyshev quadrature of the second kind cancels the angular part of the Jacobi determinant of the integral by its weights. However the weights of the Gauss-Legendre quadrature are included in all standard packages for numerical integration and thus for pure practical programming reasons, we choose Gauss-Legendre integration and the standard Chebyshev approximation, where the necessary zeros are determined by a simple formula.

of Chebyshev coefficients to be solved for. Even though after all the function $f(x)$ is approximated by a polynomial like in the case of interpolating polynomials, the Chebyshev approximation works quite different. Both being exact at the sample points of the function $f(x)$, the interpolating polynomial has an error between the sample points, which possibly can be accumulating at certain positions in the interpolation interval. The Chebyshev approximation in contrast smears out the overall approximation error over the interval under consideration.
In principle both methods, the direct angle discretisation and the Chebyshev approximation have advantages and disadvantages. Using the direct angle method one solves directly for the desired functions. If there are some properties like poles in the integration range of such functions, the direct angle method can cope with these without introducing any numerical problems besides usual matters of accuracy.
The Chebyshev approximation in contrast tends to converge very badly to the function it should approximate when there are poles or other structures likes cuts. So unless one knows that such structures do not appear, one should be careful when using the Chebyshev approximation.
On the other hand the Chebyshev approximation significantly reduces the size of the calculation, if the resulting functions are not too complicated and rapidly varying. This was the reason, Chebyshev approximations were introduced to the numerics of integral equations in the first place, when computational resources were much more scarce then they are today. In the case of BSEs it turns out that Chebyshev approximations are the best choice.
Applying the direct angle method one finds that there are no problematic structures in the angular integrals. Most importantly however the direct angle method becomes problematic for bound states with higher quantum numbers. From the angular momentum tensors in the decompositions of the Bethe-Salpeter amplitudes (see 3.3.2), higher powers of momenta arise, which lead to problems with the total grid matching, that occurs when using the direct angle method without additional interpolation in the external momentum. This grid matching leads to propagator denominators which numerically come close to zero. In principle everything is well defined and there are always terms in the numerator that cancel such zeros in the propagator denominators, when calculating

exactly. However since there appear very large numbers that are multiplied by small ones, round off errors make the direct angle method unsuitable without a vast amount of sample points. Using the Chebyshev approximations, such problems are practically absent and so we chose to employ Chebyshev polynomials.

Having discretised the BSE by the aforementioned methods, we end up with a simple linear eigenvalue problem

$$\Gamma(p_i,\rho_j) = \lambda(P^2)\cdot M(P^2,p_i,\rho_j,q_k,\theta_l)\,\Gamma(q_k,\theta_l), \tag{5.2.10}$$

where we have denoted the angle between the external and the total momentum by ρ and the corresponding angle between loop and total momentum by θ. Ordering the indices i, j and k, l respectively into a multiindex, this equations is reordered into a normal matrix equation. Depending on the goal of the calculation there are numerous methods that can be applied to such equations. In case one is interested in all bound states and corresponding Bethe-Salpeter amplitudes the most natural choice to treat this equation is the QR-method [Fra61, Fra62]. If one only searches for a single bound state in a particular region an inverse iteration with shift (Wielandt-method [Wie44]) is useful. Looking for the lowest lying bound state usually the direct iteration method is sufficient. Both the Wielandt and the direct iteration method also yield the Bethe-Salpeter amplitude, which is nothing but the eigenvector from (5.2.10) at the particular bound state mass. To obtain only the spectrum of bound state masses it is convenient to reorder (5.2.10) as

$$\left(M(P^2,p_i,\rho_j,q_k,\theta_l) - \lambda(P^2)\cdot \mathrm{Id}\right)\,\Gamma(q_k,\theta_l) = 0 \tag{5.2.11}$$

and set $\lambda = 1$. Then one calculates the determinant of $(M - \mathrm{Id})$, and looks for zeros on the negative P^2 axis, which correspond to bound state masses.

From all these possibilities, we have to choose the methods most appropriate for the task of calculating glueball properties. As already mentioned, being an initial work on the topic, we will confine ourselves to the investigation of glueball masses. The general framework, in which such a calculation can be carried out was the most important task to start with. We have developed this framework in the previous chapters (cf. 3.1, 3.2

and 3.3) dealing with the theoretical foundations. To also present an application and show where the problems in practical calculations lie, the investigation of glueball masses is a natural choice, since the mass spectrum of glueballs in pure gauge theory is known rather well.
Furthermore going deeper into the investigation might be a second step before the first, since other glueball properties than masses usually require the Bethe-Salpeter amplitude. It is simple to extract this amplitude from a BSE calculation, however whether it is meaningful or not depends crucially on the truncation. The most basic requirement on the truncation is that it reproduces the mass spectrum known from calculations using other methods. When this goal is achieved, studying other properties becomes sensible.

We will now discuss the method we apply to numerically solve the glueball BSE. To determine the masses of glueballs we will calculate the determinant of the LHS of (5.2.11) with $\lambda = 1$ for several squared bound state momenta P_B^2 and search for zeros on the negative real axis. We will do so for different quantum numbers and thus obtain a glueball mass spectrum. From the comparison with the spectrum obtained from other methods, we will be able to check the quality of our truncation scheme.
Having chosen the one-loop approximation for the *2PI*-effective action to derive our system of equations of motion for ghosts and gluons in section 3.1, we arrived at a generalised ladder-truncation for the glueball BSE. It is of course possible that a rather simple truncation like the ladder is not sufficient to describe the mass spectrum properly. However going beyond the ladder truncation, means going beyond the one-loop approximation for the effective action. The system of equations of motion for ghosts and gluons then will also include two-loop diagrams, which at present are unknown how to calculate. Since this is the first step towards a study of glueball properties using continuum methods, our truncation scheme is the most simple that respects all symmetries and avoids the aforementioned problems.

Having calculated the dressing functions of ghosts and gluons in chapter 4 and their respective continuation into the complex plane as inputs, we now have to choose the appropriate decompositions of the Bethe-Salpeter amplitudes. How such decompositions

are obtained has been analysed in detail in section 3.3. We choose the $(1/2, 1/2)$ representation for the amplitudes connected to gluons as constituents and the $(0, 0)$ representation for those connected with ghosts.
Considering states with total spin J, we find that it is not possible to construct states with parity $P = (-1)^{J+1}$ in the scalar representation for ghosts. Due to that fact, there are no ghost contributions to glueball bound states for these quantum numbers and only the diagram containing solely gluons will survive. So recalling the system of BSEs for glueballs derived in section 3.2.2 we have for states containing both ghost and gluon contributions[11]

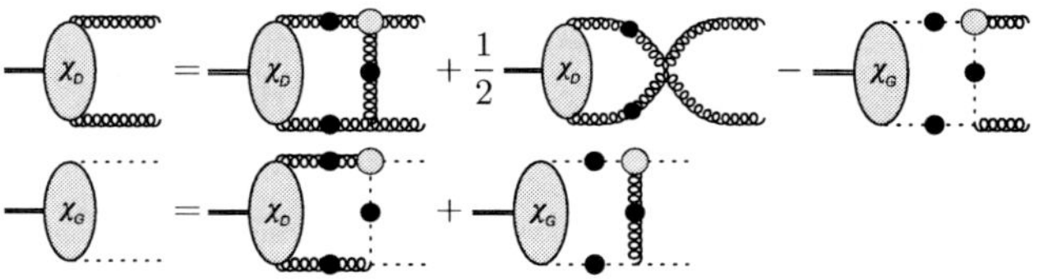

$$\chi_D = \chi_D + \frac{1}{2}\chi_D - \chi_G$$
$$\chi_G = \chi_D + \chi_G \qquad (5.2.12)$$

and for those who do not include ghost contributions

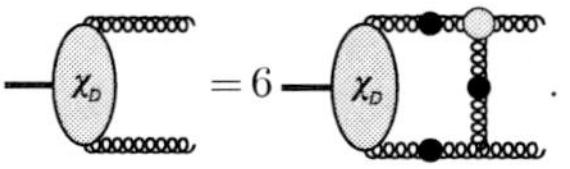

$$\chi_D = 6\,\chi_D\ . \qquad (5.2.13)$$

The only thing left is to move from Minkowski to Euclidean space, but this is trivial since the identity doesn't change at all and the Minkowski metric tensor is simply replaced by the corresponding tensor in Euclidean space namely the Kronecker tensor, while the Levi-Civita tensor also stays the same.

Having chosen states and parameters, we still have to choose an appropriate method to tackle the BSE. As already pointed out, what we actually calculate is a discretised form of the BSE, which essentially is a matrix and the desired information has to be extracted from that matrix. As mentioned before, we will do so by calculating the determinant of the BSE matrix minus the identity. This is a numerically very efficacious way to extract information from the Bethe-Salpeter matrix, since the determinant is efficiently

[11] As in equations (3.2.35) to (3.2.39) the vertex blob implies proper symmetrisation of the respective diagram.

calculated by performing an LU-factorisation on the Bethe-Salpeter matrix and then multiplying the diagonal elements. There are parallel algorithms to do so for both real and complex numbers.
In our case the only point where we have to be somewhat more careful is that due to the cutoffs, which are either rather small or large compared to one and the fact that with higher total spin, there possibly appear high powers of the loop momentum in the integral kernels, it is not sufficient to simply multiply the diagonal elements in the order they come out of the LU-decomposition. If we did, there easily can occur over- or underflows. However this is easily avoided, by ordering the set of numbers on the diagonal after the LU-decomposition with respect to their absolute values and then multiply the largest element with the smallest, the second-largest with the second-smallest on so on.

5.3 Numerical Results

We will start our discussion of the numerical study of the BSE with the results obtained using the fit-functions discussed in 4.2.3. First note that the fits were obtained by solving the Yang-Mills system, calculating the Schwinger function of the resulting gluon and fitting an analytic form to the obtained data. The combined fit was initially not done for the ghost [ADFM04]. Rather we constructed the ghost fit-function out of the definition of the running coupling. Furthermore the gluon fit-function was only constructed for scaling.
We constructed the decoupling fit-functions by analogy. For the fit-function of the scaling gluon, the Yang-Mills system has been solved for a three-gluon dressing function that differs from our construction. However since the ghost is not itself truly a fit, but a construction in case of scaling and neither are gluon and ghost fit-function fits to true data for decoupling, we will not put in the three-gluon dressing function, used to obtain the data to fit the scaling gluon in the first place, for it breaks Bose-symmetry.
Also since even the gluon is but a fit, we will have inconsistencies anyway in our calculation, whether the vertex dressing is the original one or not. We are more interested to get a taste of the general behaviour of the glueball BSE before we turn to a full

consistent numerical calculation using our solution of the Yang-Mills system in the complex plane.
The advantage of using the fits is, that they provide a means to check the impact of analytic structures on the glueball BSE, without having numerical inaccuracies included in the first place. Numerical inaccuracies in the solution of the BSE itself therefore in this case will not be remnants of inaccuracies in our input functions of ghosts and gluons. Furthermore the fits provide the possibility to study also the scaling solution, since numerically we were not able to generate solutions for the Yang-Mills system in the complex plane for this solution type.
Using the fits of course does not provide any parameters, we could use to match results to the lattice results. Of course in principle we can put in some vertex dressing providing some such parameters or vary parameters included in the fits. However doing so would further weaken the justification of the fits themselves moving them further away of their data base and essentially would be not much more than guess-work and inconsistent with the glueball BSE besides. So we will confine ourselves to simply putting the fits into the glueball BSE and calculate a results "as is". To get an impression of how well our glueball BSE works and what difficulties we might have to expect doing so is fair enough.

Let us start with the decoupling type of solutions. In previous sections (cf. sections 3.2.2, 4.2, 4.2.2, 4.2.3 and 5.2) we have extensively discussed all foundations and subtleties of the glueball BSE, so we will not restate the framework here again, but confine ourselves to the presentation of results.
We will show the results of scans of the determinant of the glueball Bethe-Salpeter matrix minus the identity. These are conducted along the axis of real squared momenta. The most interesting region is of course the negative real axis, which corresponds to time-like Minkowski momenta, where physical bound states can be found. We will do so for a set of states with $J^{PC} = 0^{++}$, 0^{-+}, 2^{++}, 2^{-+}, which should be enough get an impression of the numerical behaviour of the Glueball BSEs. Looking at the determinant of the BSE matrix minus the identity means that zero crossings indicate bound states.
Figures 5.1, and 5.2 show the results for the decoupling fit. In general we observe a oscillatory behaviour, especially in the range between $0\,\mathrm{GeV}^2$ and $-10\,\mathrm{GeV}^2$, while on

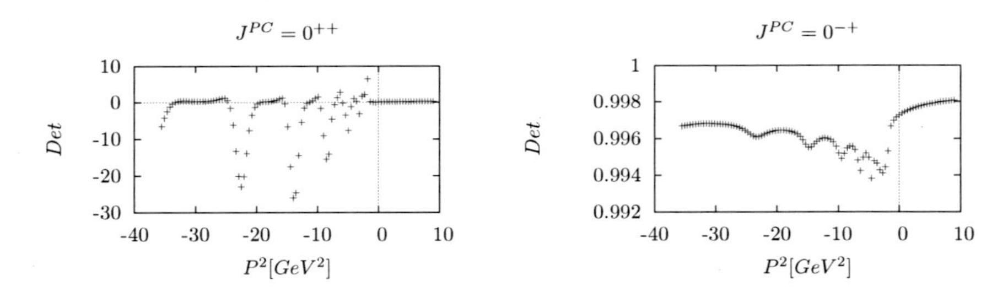

Figure 5.1: The determinants of the scalar (left) and pseudoscalar (right) glueball, calculated using the decoupling fit-functions for ghosts and gluons. The results in general seem to oscillate and exhibit noisy behaviour above $-10\,\mathrm{GeV}^2$. The pseudoscalar state is varying only mildly, indicating only small interaction strengths being present.

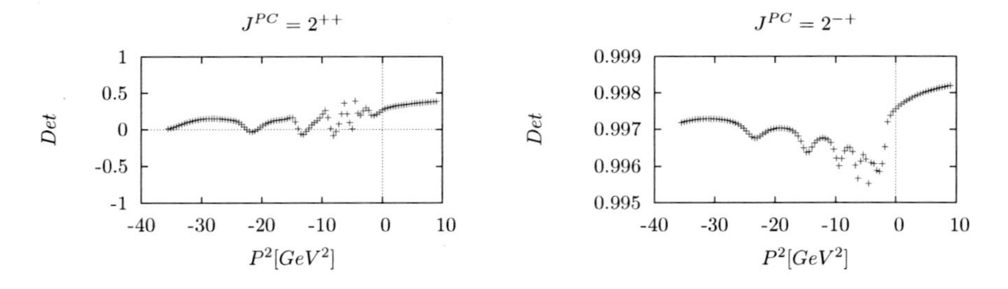

Figure 5.2: The determinants of the spin two tensor (left) and pseudotensor (right) glueball, calculated using decoupling fit-functions. As for scalar and pseudoscalar states we observe oscillatory behaviour and noise above $-10\,\mathrm{GeV}^2$. Again in the parity-odd state, there seems to be only small interaction strength present.

the positive real axis the behaviour abruptly becomes smooth and monotonous. These oscillations are also seen for states with odd parity. These states are furthermore far from generating any bound state. The scalar and the spin two tensor do exhibit bound states, though we do not assign to them much significance, due to the aforementioned inconsistencies. Furthermore we note, that the variation of the determinants calculated for odd parity states is greatly reduced compared to the parity-even states. Since the only significant difference, which appears to be important enough to be the cause for that, is the fact that there are no ghost diagrams contributing to the glueball BSE for parity-odd states, it seems sensible to attribute the greatly reduced variation of the determinant to that absence of ghost contributions. Turning to the scaling type of solution we look at figures 5.3 and 5.4. We note that in general the behaviour is again oscillatory. However the rate of variation is greatly enhanced compared to the results using the decoupling type fits. This is also true for the parity-odd states, although they also are varying comparatively mildly. Also as in the decoupling case we observe considerable oscillations in the region between $0\,\mathrm{GeV}^2$ and $-10\,\mathrm{GeV}^2$. Since we observe that pattern in both the results obtained using the scaling and the decoupling type fits, which are themselves free of numerical noise, we conclude that this is a feature of our glueball BSEs. The exact reason for this noise is not clear yet, but one might suspect it is connected to the influence of the poles, that are present in the fit-functions. Again we refrain from attributing much significance to zeros found for the parity-even states for the aforementioned reasons. We conclude that, looking at the results obtained from the glueball BSE, we gain three insights.

Firstly, we can generated bound states from the glueball BSE, at least for the parity-even states. The parity-odd states do not seem to gain enough interaction strength from the purely gluonic diagram to provide for the formation of bound states.

Secondly we see that for some reason, generally the result might exhibit noisy behaviour, in the region between $0\,\mathrm{GeV}^2$ and $-10\,\mathrm{GeV}^2$, possibly induced by analytic structures of the fits. We have to expect numerical noise in that region for the full numerical calculation also.

Thirdly the comparison of the results obtained with decoupling and scaling fits indicates that the effect of switching between these two types of solution mainly results in a significant enhancement of the variation of the solution. Whether or not this will have a significant influence on the position of bound states is unclear. Since the lowest zero transitions are generated in the noisy region, we cannot make a solid statement.

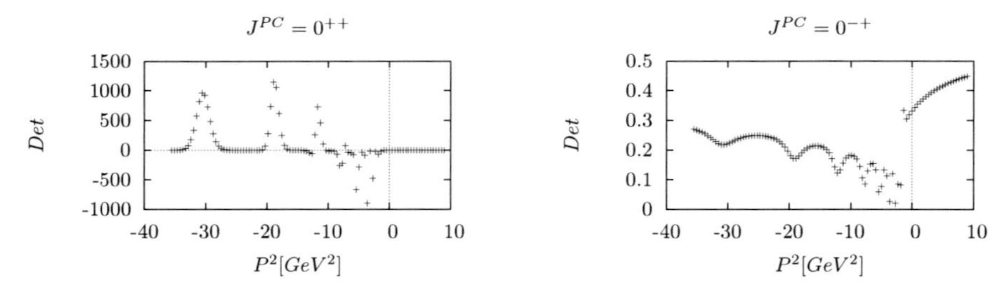

Figure 5.3: The determinants of the scalar (left) and pseudoscalar (right) glueball, using the scaling fit-functions for ghosts and gluons. The results in general seem to oscillate and exhibit noisy behaviour around $-10\,\mathrm{GeV}^2$. However the magnitudes are greatly enhanced compared to the decoupling solutions. The pseudoscalar state again is varying only mildly, indicating only small interaction strengths being present.

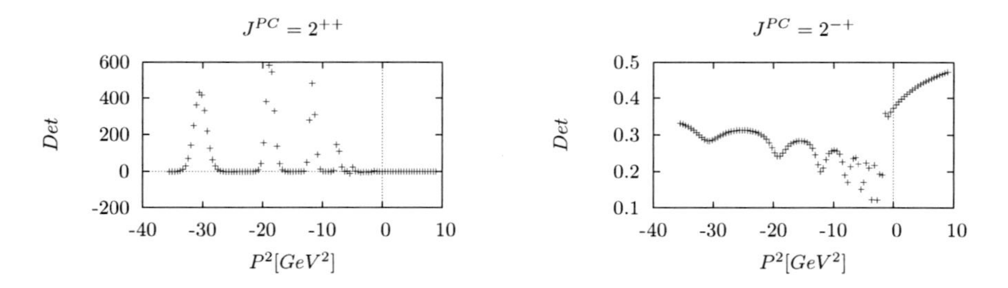

Figure 5.4: The determinants of the spin two tensor (left) and pseudotensor (right) glueball using decoupling fit-functions. As for scalar and pseudoscalar states we observe oscillatory behaviour and noise around $-10\,\mathrm{GeV}^2$. Also as before, we find that the magnitudes are greatly enhanced compared to the decoupling results. Again in the parity-odd state, there seems to be only small interaction strengths present.

As the final part of this thesis, we will now turn to the full and consistent numerical solution of the glueball BSE, using the numerical solutions for ghost and gluon dressing functions obtained for complex momenta in section 4.2. As the truncation scheme is consistent for both the system of SDEs for ghosts and gluons and the glueball BSE, we will expect that bound states we observe are significant predictions of glueball masses. These masses we will compare to the corresponding findings in the literature. In section 4.1, we have discussed the model dressing of the three-gluon vertex, that we use in our SDE and BSE calculations. For convenience of the reader we will restate it here

$$
\begin{aligned}
{}^{(A^3)}\Gamma(x,y,z) = a\,\mathfrak{Re}\Bigg[&\left(1+\omega\log\left(\frac{x}{\mu^2}\right)\right)^{-\gamma} \\
&\left(1+\omega\log\left(\frac{y}{\mu^2}\right)\right)^{-\gamma}\left(1+\omega\log\left(\frac{z}{\mu^2}\right)\right)^{-\gamma}\Bigg].
\end{aligned}
$$

We have chosen the set of parameters $\omega = 1$, $\mu^2 = 1$ along with the anomalous dimension of the gluon $\gamma = -13/22$. That leaves the parameter a for variation. We have varied the parameter from $a = 1.0$ to $a = 2.0$.[12] We found that the variation of a does not give a significant change of the results, neither for the solutions of the Yang-Mills system nor for the glueballs. We thus choose to fix $a = 1.0$ and use a different parameter to fit such that the glueball mass spectrum comes out well.

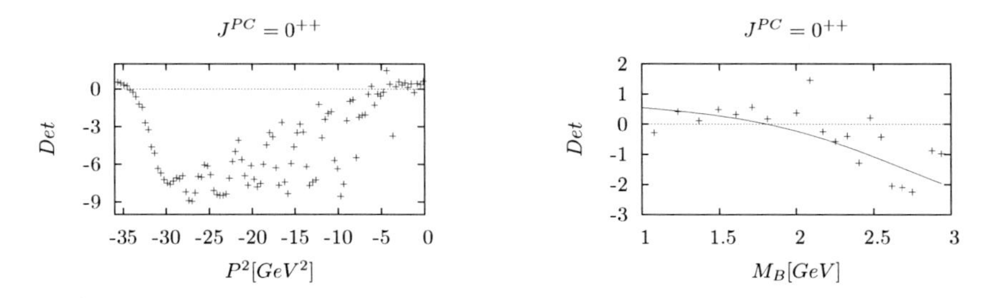

Figure 5.5: The determinant of the scalar glueball, as found in the full consistent numerical calculation. The left panel shows the full range scan (in GeV^2) and the right shows a magnification around the location of the lowest bound state (in GeV).

[12] Recall that the range in which we can vary this parameter is limited due to the stiffness of the Yang-Mills system cf. 4.1.

Up to now we have always fixed the scale emerging dynamically solving the Yang-Mills system, to the scale found in gauge-fixed lattice calculations (cf. 4.1 and 4.2). However we already pointed out that there are differences between fixing to Landau-gauge on the lattice and in continuum field theory (cf. 4.1). This in turn means that the scale Λ_{lat} found in such lattice calculations is not necessarily exactly what would be found in an exact continuum calculation. We choose to use a modified scale $\Lambda_{mod} = \zeta\, \Lambda_{lat}$ to rescale the internal units, that emerge in our solution of the Yang-Mills system, to physical ones. Thereby we obtain a parameter ζ, which we fit such that we obtain a glueball mass spectrum consistent with the literature. We find that we do not need a large shift ζ. We obtain a good agreement of our results using $\zeta = 2$.

We have conducted our calculations for a variety of states, including "unnatural" quantum numbers like $1^{++}, 1^{--}$ and 3^{--} for completeness. In principle the states with negative charge-conjugation parity are not attributable to bound states of two gluons as pointed out before (cf. 5.1). However flipping charge-conjugation parity is done by merely multiplying the amplitude with a factor $(P \cdot q)$ (see 3.3) and does not change the result much in the ladder truncation.[13] For negative-charge parity however we have results to compare and see the impact of the fact that the corresponding state cannot be a two-gluon bound state. Comparing the result for the vector glueball will clearly show the inconsistency of such a state with other calculations, where the vector state is (or can be) comprised of more than two gluons.

We start the presentation of our numerical result with the scalar and spin two tensor glueballs. We are only interested in the lowest bound states. Although we do also obtain candidates for excited states, we will not consider them as such, since our truncation scheme is very simple and not suited to reproduces complicated features like excited states. The results are mostly fuzzy for excited states and if not then these states are totally inconsistent with corresponding states given in the literature.

[13] The same is true for corresponding calculations of mesons.

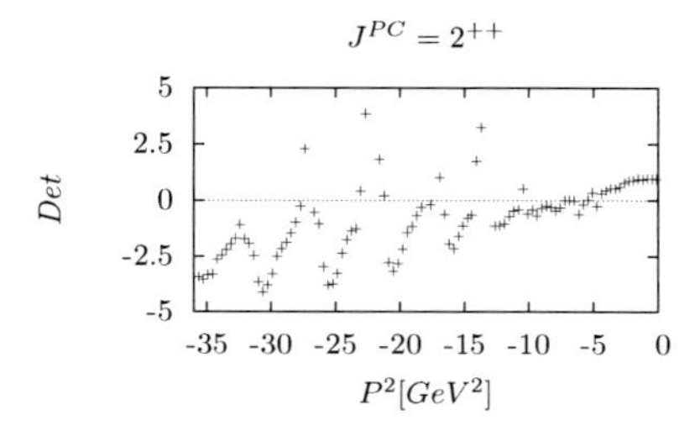

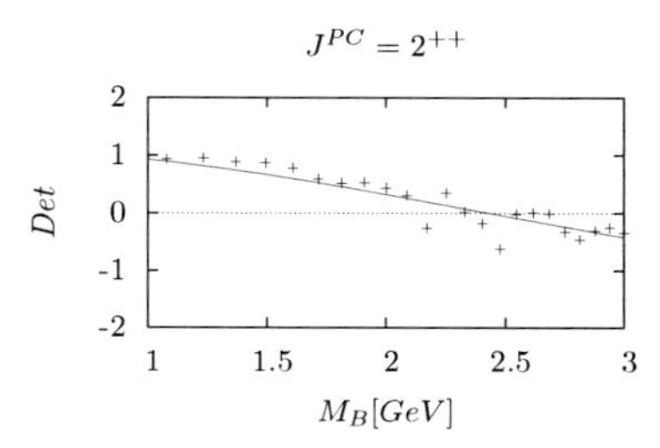

Figure 5.6: The determinant of the spin two tensor glueball, as found in the full consistent numerical calculation. The left panel shows the full range scan (in GeV^2) and the right shows a magnification around the location of the lowest bound state (in GeV).

For all states we scanned along the negative real axis on the interval $[-36\,\mathrm{GeV}^2 : 0\,\mathrm{GeV}^2]$ for the squared total momentum P^2 correspondingly to a range 0 GeV to 6 GeV of bound state masses. When there was a clear signal of a bound state, we chose to give a second plot, where we magnify the region around the bound state. We have determined the position of the bound states using smoothing splines [CW79] for all states except the 4^{++}, where it is obviously better to use normal splines, since it is rather smoothly behaved around the bound state. The smoothing splines are useful, because the position of the bound states lies within the region, where the determinant is affected by numerical noise. It is the same region from $0\,\mathrm{GeV}^2$ to about $-10\,\mathrm{GeV}^2$, which we found being noisy using the fit-functions before. we conclude that this is a general feature of the glueball BSE. We suspect that this is a numerical problem, possibly connected with round off errors due to rather delicate cancellations in the integration kernels. Since it does not vanish when states do not have contributions from ghosts, these cancellation issues are a general feature of all integration kernels of the BSEs.

The determinants obtained for the scalar and pseudoscalar glueballs are shown in figures 5.5 and 5.6. In the case of the scalar, the region where the lowest bound state is located, is quite noisy. To apply a smoothing spline we had to remove one point located around $-4\,\mathrm{GeV}^2$ that is clearly a runaway and which rendered the smoothing spline awkward.

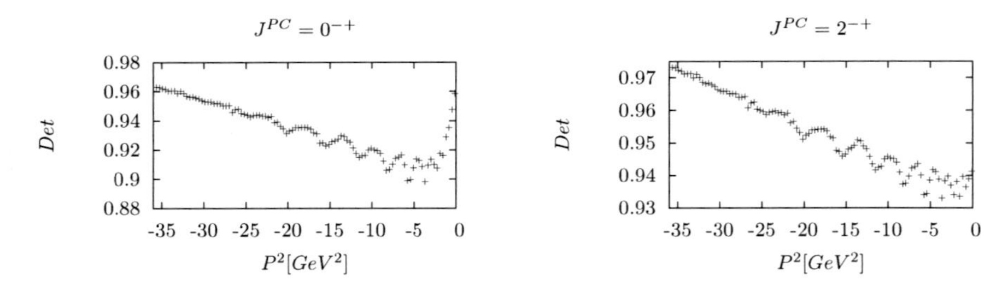

Figure 5.7: The determinants of the pseudoscalar (left) and spin two pseudotensor (right) glueball, as found in the full consistent numerical calculation. In both cases we do not observe any signals of bound states. The variation of the determinants is quite small.

The smoothing spline is shown in the right panel of 5.5. We find the bound state mass of the scalar at 1.81 GeV.[14] We also observe a signal of an excited state in the 0^{++} channel. It is even remarkably clear. However its mass of almost 6 GeV lets us discard that signal.

For the spin two tensor shown in 5.6 the situation is similar as for the scalar glueball. However around the lowest bound state, we find less noise. We locate the mass of the 2^{++} glueball ground state using a smoothing spline at 2.42 GeV. There are several signals of excited states, however they seem to be noisy and unreliable. The clearest signal in a sense, would be located at a mass of about 4.5 GeV, which is very large. We thus also discard all these possible excited state signals.

We now turn to the spin zero and two states with odd parity. In figure 5.7 we have shown the full scan of the respective determinants. In contrast to the corresponding parity even states, we do not find any bound state signals. In both cases the variation of the determinants is quite small. This is in agreement with the results of the corresponding states we obtained before using the fit-functions of both types. In figure 5.8 we have

[14] One should not take the display of two significant figures too seriously. Clearly the errors, that effect our calculation do not really allow for such an accuracy. However we nevertheless write them down in order to be consistent with results given in 5.2 to 5.6.

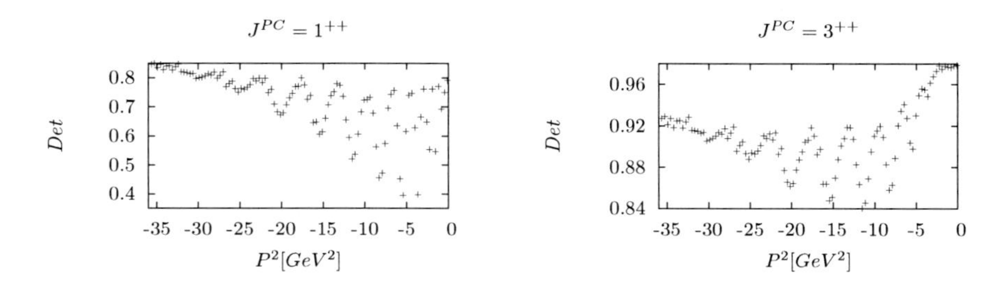

Figure 5.8: The determinants of the axialvector (left) and spin three pseudotensor (right) glueball, as found in the full consistent numerical calculation. In both cases we do not observe any signals of bound states. The variation of the determinants is quite small.

plotted the determinants of the states 1^{++} and 3^{++}. Both also only include gluonic contributions and again we observe no bound states and only mild variations for the pseudotensor. The determinant of the axialvector varies significantly stronger. However this state is forbidden by Yang's theorem, which could possibly mean that this amplitude is also meaningless in our approach. Still in all states comprised solely of gluons we have considered so far, we see the same picture. The interaction strength seems to be significantly reduced, leading to only mild variations of the determinant, yielding no signals of bound states. This further confirms our suspicion that this feature might be connected to the missing contributions of ghosts in these channels.

Turning to the spin four tensor, whose determinant is shown in 5.9, we find the lowest lying bound state outside the noisy region. We thus chose not to apply a smoothing spline and to use an ordinary spline instead. The lowest lying bound state is found at a mass of 4.03 GeV.

Beyond the lowest lying zero we find oscillations leading to the first excited state candidate close by the ground state, which is inconsistent with findings in other studies, so we again discard these states as artifacts. Note that the 4^{++} state also gets a

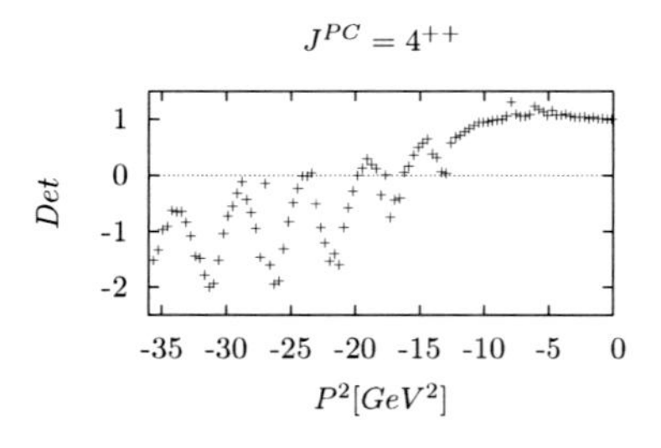

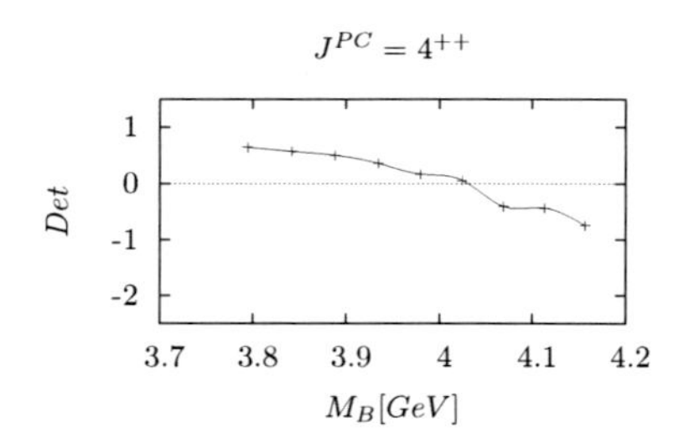

Figure 5.9: The determinant of the spin two tensor glueball, as found in the full consistent numerical calculation. The left panel shows the full range scan (in GeV^2) and the right shows a magnification around the location of the lowest bound state (in GeV).

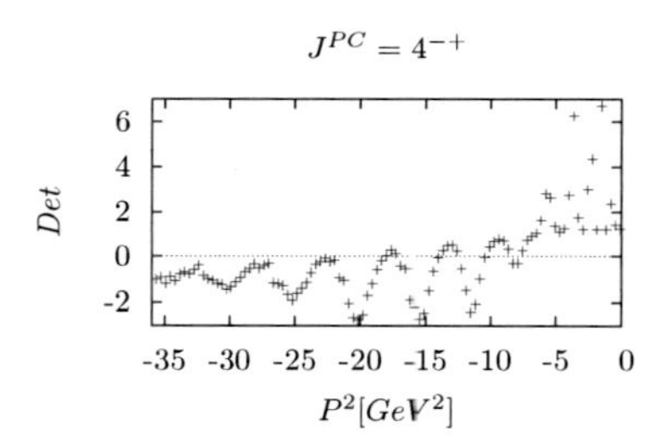

Figure 5.10: The determinant of the spin two tensor glueball, as found in the full consistent numerical calculation.

contribution from the basic spin two amplitude (3.3.12), which means that it is described by three amplitudes (cf. 3.3.1).

The next state we have studied is the spin four pseudotensor. It is another state that gets only contributions from the purely gluonic diagram in the glueball BSE. However as can be seen in figure 5.10, we find a bound state in contrast to the purely gluonic states considered before. The bound state is located at a strangely low mass of about 2.8 GeV, which is not only inconsistent with literature but also with our previous

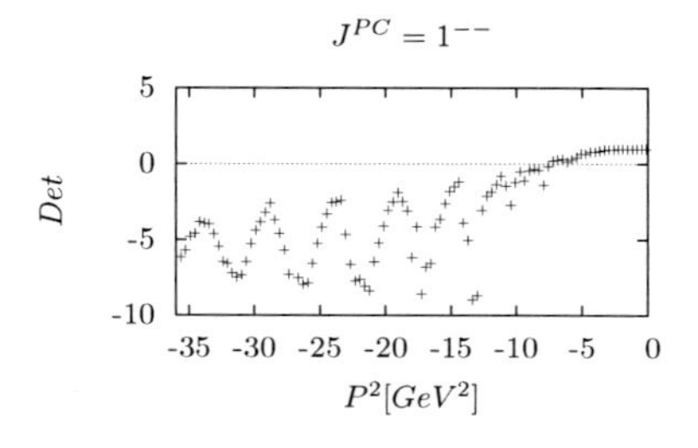

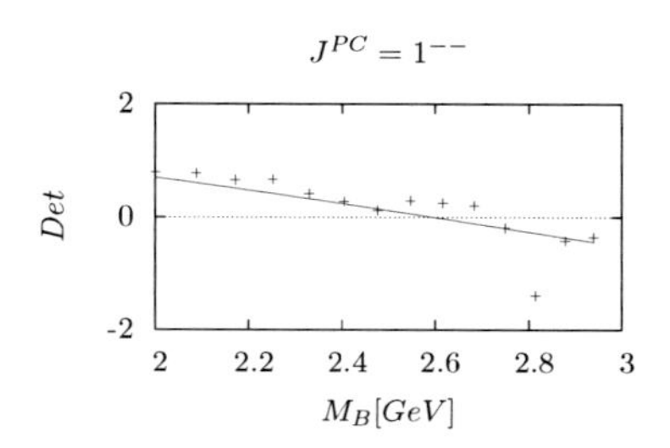

Figure 5.11: The determinant of the vector glueball, as found in the full consistent numerical calculation. The left panel shows the full range scan (in GeV^2) and the right shows a magnification around the location of the lowest bound state (in GeV).

findings. Considering that up to now all purely gluonic amplitudes clearly had no bound states and the extremely low position of the lowest bound state in the 4^{-+}, we refrain from accepting this as a valid result and discard it. Also again there are lots of signals of excited states even close by the ground state. We conclude that we see nothing but oscillations in the determinant maybe of numerical nature or artifacts of our truncation. We have furthermore looked at two states, which are not constructible in terms of direct products of vector representations of gluons (see 5.1). The first we will discuss is the vector glueball, which is forbidden by Yang's theorem. In figure 5.11, we have plotted the determinant found for that channel. As can be seen, we find a bound state with a mass of 2.59 GeV. Compared to the 2^{++} glueball it is remarkably heavy. However compared to the findings in all in all other approaches it is extremely light. So we can conclude that the 1^{--} ground state obtained in our approach, which definitely is comprised of two gluons is inconsistent with the literature. Especially lattice calculations, where the number of gluons comprising a bound state is not fixed, indicate that the mass we find is unreasonably low, thus confirming Yang's theorem.

The last state we will consider has quantum numbers 3^{--}. It is described by three amplitudes, since as for 2^{++} and 4^{++} states, there contributes an amplitude build from the basic spin two invariant (3.3.12).

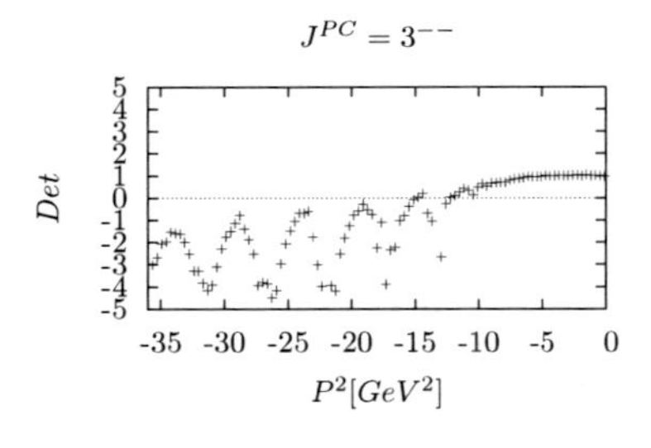

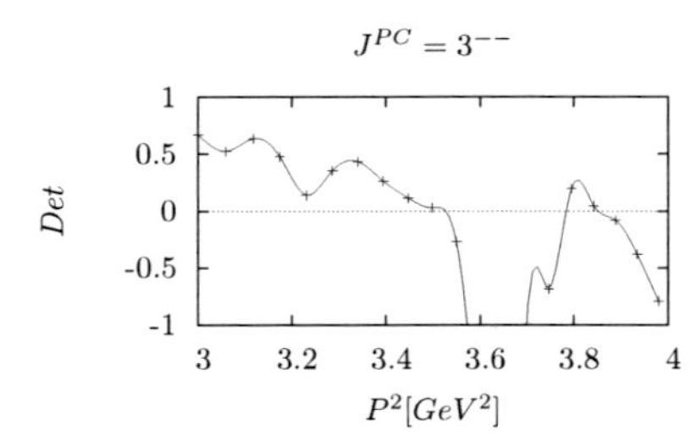

Figure 5.12: The determinant of the spin three tensor glueball, as found in the full consistent numerical calculation. The left panel shows the full range scan (in GeV^2) and the right shows a magnification around the location of the lowest bound state (in GeV).

As we did for the vector glueball, we find a low lying bound state, the mass of which is 3.53 GeV. This state again is quite definitely excluded from the two-gluon bound state spectrum, since it has odd charge parity. We calculated it only because, we have data to compare with. However the charge parity flipped state 3^{-+} will likely have a quite similar mass. It is however not "as forbidden" as the 3^{--}, since it has even charge parity as two-gluon systems have to. It is forbidden only by helicity counting in the representation using direct products of gluon vector representations. As already discussed in section 5.1, in a fully covariant approach we can construct such a state and the mass of the 3^{--} state, we found here indicates that in the glueball mass spectrum there might be the possibility of an "exotic" glueball state with quantum numbers 3^{-+} in the mass range, where other spin three states are found. Having discussed in detail the determinants found solving the glueball BSE(s) for various states, we will now collect the result and compare them to the findings in studies using other methods as given in tables 5.2 to 5.6. Firstly in table 5.7 we compare our findings in channels with natural quantum numbers for two-gluon systems. As can be seen the results agree very well. We conclude that our method can reproduce the mass spectrum found in the literature well, except for the states that do not have any contributions coming from diagrams in the BSE containing ghosts. We suspect that in these channels the interaction strength, solely provided by the gluonic diagram, is too low. Physically that means that the state becomes very heavy,

	masses (GeV)						
J^{PC}	lattice		Hamiltonian/ Regge theory		constituent models		our result
0^{++}	1.71	[CAD+06]	1.98	[SS03]	1.71	[Bui07]	1.81
	1.55	[B+93]	1.58	[KS00]	1.60	[MSSB08]	
	1.75	[LWC02]			1.86	[MSSB08]	
	1.73	[MP99]					
2^{++}	2.39	[CAD+06]	2.42	[SS03]	2.53	[Bui07]	2.42
	2.23	[B+93]	2.59	[KS00]	2.05	[MSSB08]	
	2.31	[B+93]			2.38	[MSSB08]	
	2.42	[LWC02]					
	2.40	[MP99]					
4^{++}	3.65	[LW02]	3.99	[SS03]	3.60	[Bui07]	4.03

Table 5.7: Comparison of our numerical results for masses of glueballs having natural quantum numbers with other studies.

possibly infinitely heavy, which means it disappears. This on the other hand leads us to the speculation that the anomalous mass hierarchy between the states 0^{++}, 2^{++} and 0^{-+} in Landau gauge is due to the absence of ghost contributions in the 0^{-+} channel, which presumably increases its mass. However until our truncation scheme, which is the simplest possible preserving ultraviolet behaviour of ghosts and gluons known from perturbation theory, as well as Bose-symmetry, is improved such that the interaction strength in the purely gluonic channels is sufficient for the formation of bound states, we can only speculate about the reason for the mass anomaly.

In table 5.8, we show the results of our calculations in channels with unnatural quantum numbers. We did this because in contrast to representations that use direct products of gluon helicities, in our fully covariant approach, we can construct amplitudes for bound states of two gluons with such quantum numbers. Both states are quite definitely forbidden for bound states of two gluons. Both because they have odd charge-conjugation parity and the vector because of Yang's theorem. Nevertheless we

	masses (GeV)			
J^{PC}	lattice	Hamiltonian/ Regge theory	constituent models	our result
1^{--}	3.83 [CAD+06]	3.49 [KS00]	3.43 [MSSB08]	2.59
	4.36 [B+93]		4.00 [MSSB08]	
	3.85 [MP99]			
3^{--}	4.20 [CAD+06]	4.03 [KS00]	3.57 [MSSB08]	3.53
	4.13 [MP99]		4.17 [MSSB08]	

Table 5.8: Comparison of our numerical results for glueball masses of unnatural states with other studies.

can construct these amplitudes and use them in BSEs. We find that the vector is much lighter than what is found in the literature. Especially lattice calculations, where the number of gluons comprising the state is undefined, provide a good benchmark. Being so far away from the masses found in the literature, we can conclude that there is no state hidden in this sector that for some reason eludes Yang's theorem. In case of the spin three tensor matters possibly can be a little more subtle. The state we calculated definitely is excluded from the spectrum of two gluon bound states, because of its odd charge parity. However in the ladder truncation flipping the parity usually does not make much difference. We chose to calculate the state 3^{--}, because we have numbers to compare our findings for its mass against. From helicity counting both states 3^{--} and 3^{-+} are forbidden. However since in our approach they can be constructed and the resulting mass is quite in the range where spin three gluon bound states presumably can be found, we conclude that the existence of an *exotic* glueball state having spin three and odd parity cannot be excluded so far.

Chapter 6

Summary, conclusions and an outlook

In this thesis we have investigated the Bethe-Salpeter equation for glueballs. The main goal was to obtain a spectrum of glueball masses in a consistent way.
The mass spectrum of glueballs has been investigated before in various studies [B+93, SS03, GSG+08, LECdAB+02, LEBC06, NRW01, LWC02, LW02, CAD+06]. The most reliable results come from lattice gauge theory. However in lattice gauge theory the mechanisms of bound state formation are obscured by averaging over field operators and thus hard to access. Also unquenching is computationally very costly. In a continuum approach these problems are rather easily circumvented, which is an important advantage.
In recent years much effort was spent investigating non-perturbative QCD in the continuum e.g. [MR03, vSHA98, FA03, AvS01, BLLY+08]. The main topics in these studies have been the fundamental degrees of freedom i.e. quarks, gluons and (depending on the gauge) ghosts.[1] Since none of these degrees of freedom is observable, it is important to connect the results of these investigations of fundamental degrees of freedom of QCD to

[1] Most investigations have been done in Landau gauge. There are also studies done in Coulomb gauge (e.g. [WR08] and references therein) and in maximally abelian gauge (e.g. [HSA10] and references therein).

observables. For quarks, this has been done very successfully in a number of studies (e.g. [AWW02, MR03, MT00, FW09] and references therein). For gluons this has not been so. In this thesis we make the first attempt to conduct such an application.
A convenient starting point for the investigation of the basic degrees of freedom of continuum QCD beyond perturbation theory are the corresponding Schwinger-Dyson equations. From these, information about the non-perturbative properties of the degrees of freedom can be obtained in form of dressing functions [RW94, AvS01]. Still these results are connected to fields that are not observable themselves. The fields therefore have to be used in some way to calculate observable quantities. A way to do this, that has been used successfully in a number of studies is to combine results from Schwinger-Dyson equations and the Bethe-Salpeter equation. This combined Schwinger-Dyson/Bethe-Salpeter approach proved successful for quark-meson studies, providing meson spectra and electromagnetic form-factors (e.g. [MT00, MR03, WR08] and references therein). Also the combination of Schwinger-Dyson equations with the Faddeev equation describing the properties of bound systems of three particles, was successfully used to study baryons (e.g. [EAKN10, Eic11, SAEVCA11]). It is thus quite natural to use an Schwinger-Dyson/Bethe-Salpeter approach to connect results obtained for gluons and ghosts with observable states i.e. glueballs. This was the main idea of this thesis.
Starting from the knowledge about the Schwinger-Dyson/Bethe-Salpeter approach to mesons and the solutions for ghost and gluon dressing functions for positive real squared momenta, in this thesis several fundamental problems have been tackled and solved.
The Bethe-Salpeter equation hitherto was solved only for bound states of a single type, that do not mix with any other. Studying bound states of gluons in Landau gauge, the situation in this work was different. Since in Landau gauge there are degrees of freedom of the gluon, which are separated from the gluon field and encoded in another, namely the ghost field, the necessity arises to also consider bound states that are connected to these ghosts fields. However this leads to a set of coupled Bethe-Salpeter equations of bound states mixing. We presented a rigorous derivation of such a set of Bethe-Salpeter equations, which has to our knowledge not been done before. Furthermore the derivation is very general, easily allowing for the generation of Bethe-Salpeter equations in

approximations beyond the ladder truncation. The Bethe-Salpeter equation(s) thus obtained are consistent with the Schwinger-Dyson equations of the constituents, which are automatically derived in the process of deriving the Bethe-Salpeter equations. Also the procedure does not break any symmetry included in its starting point, namely the *2PI* effective action.

Another necessary ingredient of a study of the Bethe-Salpeter equation(s) for glueballs were the amplitudes corresponding to a given set of quantum numbers. Since Bethe-Salpeter equations have been studied for a long time, there has also been done a lot of work concerning the form of suitable amplitudes. Mostly however these have been given for non-relativistic limits, allowing for wave decompositions of the amplitude (e.g. [KUB+96]). Instead we were interested in fully covariant representations. In case of quarks these have been given in a number of places mostly for scalar or vector particles and rarely for spin two tensors with even parity (e.g. [LS69, AWW02] and references therein).

We have extended this to an exhaustive account of all quantum numbers accessible for Bethe-Salpeter amplitudes for a wide range of representations of the constituent particles including those needed for Bethe-Salpeter studies of glueballs.

Such a complete account has to our knowledge not been given in the literature before.

The next necessary ingredient into our investigation of glueballs using a Bethe-Salpeter approach, was the solution of the system of coupled Schwinger-Dyson equations for ghosts and gluons for complex momenta. This system has been studied before, but only for positive momenta. Since an (at present virtually unavoidable) technique for solving the Bethe-Salpeter equation is the Wick-rotation, solutions for the dressing function of ghosts and gluons are needed for complex squared momenta. Thus we had to solve the coupled system of ghosts and gluon Schwinger-Dyson equations in the complex plane. Solutions in the complex plane have been found before for the quark dressing function using model gluons as exchange particles[MT00, AWW02, FNW09]. This has been done mostly in the ladder approximation, where it was easily possible to employ a model gluon as exchange particle. Then the solution of the quark Schwinger-Dyson equation could be obtained rather easily by adjusting the momentum flow in the diagrams of the Schwinger-Dyson equation such that only the model gluon was to be known in the presence of imaginary parts in the momentum. However going beyond the ladder truncation gave rise to the necessity to solve the quark Schwinger-Dyson equation directly for complex quarks. The method applied to do this is called the *shell-method* [FNW09] and we used it to solve the coupled system of gluon and ghost Schwinger-Dyson equations in the complex plane.

There has been a study of those dressing functions before, which was aimed on the analytic structure of the gluon propagator in general [ADFM04]. This was done to show that the gluon propagator obtained from the coupled system of ghost and gluon Schwinger-Dyson equations violates positivity, which is a signal for confinement. To do this the authors constructed an analytic fit to the data of the gluon propagator obtained from the coupled system of Schwinger-Dyson equations and its corresponding Schwinger function. We took the analytic fit of the gluon propagator (which was of scaling type) and constructed a corresponding ghost propagator fit, using a non-perturbative definition of the running coupling of pure gauge QCD. By analogy we also constructed two such fit functions for decoupling type solutions. We compared these to our numerical results of the dressing functions of ghosts and gluons for complex momenta and found that the results are quite similar for a wide range of momenta and that the differences are not very large (though possibly significant).
Finally we put all peaces together and solved the Bethe-Salpeter equation(s) for glueballs numerically. We did so for the fit-functions of ghost and gluon dressing functions mentioned before to gain an insight of the behaviour of the glueball Bethe-Salpeter equation(s), without the influence of numerical inaccuracies coming from the input functions. Then we solved the full consistent glueball Bethe-Salpeter equation(s) using the ghost and gluon dressing functions obtained numerically from the coupled system of Schwinger-Dyson equation of ghosts and gluons. Doing so we obtained glueball masses, which we compared to results from previous studies using different approaches.
While for the fit-functions we did not find convincing results indicating bound states, we did so in the full and consistent numerical treatment of the glueball Bethe-Salpeter equation. We found that there are no bound states with spin J and parity $P = (-1)^{J+1}$. The reason for this is likely to be found in our truncation scheme. These states do not mix with states comprised of ghosts, for there is no way to construct the corresponding

amplitudes. Thus only purely gluonic interactions contribute to these states, which means that only a single diagram is present and quite possibly this does not provide sufficient integrated strength of interaction for the generation of low-lying bound states. The other states in contrast did generate low-lying bound states. We found good agreement of the corresponding groud state masses with results found in the literature for natural quantum numbers of bound states of two gluons. Furthermore we calculated masses for unnatural states and found that Yang's theorem is at work in a fully covariant approach also.

We conclude that in this work we have shown that the combined approach using Schwinger-Dyson and Bethe-Salpeter equations to study glueballs, is a viable method to investigate glueball properties in continuum field theory. We used the supposedly most simple truncation scheme that provides a consistent set of Schwinger-Dyson and Bethe-Salpeter equations and preserves all necessary symmetries. We presented a quite general derivation of a system of Bethe-Salpeter equations suited for the description of glueballs as well as a version which already includes unquenching and glueball/meson mixing effects. In general the derivation is applicable to systems where bound states of different types mix. Even though being very simple the truncation scheme allowed for the generation of bound states for quantum numbers that include ghost contributions. The results for the corresponding ground state masses are in nice agreement with predictions from other approaches. This indicates that the concept as a whole is indeed suitable for the description of glueballs.
Furthermore with this treatment of the glueball Bethe-Salpeter equation we presented the first consistent connection of ghost and gluon dressing functions obtained from Schwinger-Dyson studies to observables. The comparison of our Bethe-Salpeter results with the literature indicates that the Schwinger-Dyson results for ghosts and gluons are valid and applicable to the construction of observables.
The results for ghost and gluon dressing function presented here are however not yet of a quality good enough to use them for further investigations about their analytic structure, which would possibly provide deep insight in the mechanisms of confinement and dynamical mass generation in the gauge sector of QCD. However maybe this is not

much more than a problem of size of the numerical calculation or the employed numerical method.
By and large we conclude, that our goal, providing a first consistent study of glueballs in the combined Schwinger-Dyson/Bethe-Salpeter approach was successful and provides a reliable foundation for future investigations.

Having presented the foundations of the combined Schwinger-Dyson/Bethe-Salpeter approach to glueballs, we are now in the position to consider future perspectives.
Firstly the truncation scheme to employ has to be more sophisticated than the simple one used in this thesis. We found that seemingly our truncation scheme does not provide for enough interaction strength for the formation of light bound states including solely gluonic contributions. This leads us to suspect that a suitable truncation scheme necessarily has to go beyond the ladder truncation, or maybe more precisely beyond what we called "generalised ladder" truncation.

A possible alternative would be a better dressing of the three-gluon vertex. However this dressing is more or less fixed in its behaviour for high-momenta. There can be significant changes in the infrared behaviour, but the formation of glueballs is hardly sensitive to such. What is left would be the intermediate momentum range, which is also somewhat fixed being the transition range between high and low momentum behaviour. Thus a change in the low momentum behaviour can possibly only influence the formation of bound states indirectly via this transition region. Whether this is sufficient to allow for the formation of bound states without ghost contributions is unclear. Thus it might also be necessary to use a more complex vertex parametrisation, such as the one given in [BC80], where the each component has its own dressing function. This would be a pure gauge QCD analog to going beyond the rainbow truncation in the quark sector.

The first real goal for the future is of course to predict a physical glueball spectrum. Since glueballs naturally mix with mesons, this entails that unquenching effects have to be taken into account. We provided a derivation of sets of Bethe-Salpeter equations capable to allow for such calculations. Even though we found that the

simple generalised ladder truncation might not be sophisticated enough and has to be consistently extended. Still including unquenching and meson/glueball mixing in our approach seems tractable in the near future. Such a study would provide the first realistic prediction of glueball masses in full QCD, which naturally would be of decided interest for experimentalists.
It might well be expected that in such an extension the purely gluonic states will still be unbound. Here the next step would also be to include more complicated vertices and diagrams to go beyond the ladder (and/or rainbow) truncation, as has already been done in the quark sector.
The foundations we presented also can be fairly easily adapted to Faddeev equations, which allow for the investigation of oddballs (bound states of three gluons). Furthermore the solution of the system of ghost and gluon Schwinger-Dyson equation in the complex plane seems refinable in the near future to provide reliable information of the analytic structure of the dressing functions of ghosts and gluons, which is highly interesting from the view point of fundamental local field theory, since it will provide viable information in the search for confinement and dynamical mass generation mechanisms.
To extend the truncation scheme beyond the ladder truncation will necessarily lead to a system of Schwinger-Dyson equations for ghosts and gluons that include the two-loop diagrams, we omitted. It is still unclear how to solve that system even for real and positive momenta [Blo03]. To remedy this probably is the another step to be taken in the near future. From comparisons of the ghost and gluon Schwinger-Dyson equation with analogous equations derived from the functional renormalisation group [FMP09] and from studies replacing the two-loop diagrams by effective one-loop approximations [Blo03], it is known that the impact of these two-loop diagrams on the solution, especially of the gluon dressing function, is an enhancement of the functions at intermediate momenta leading to results closer to those obtained in gauge-fixed lattice theory. However this is the region, where bound state formation takes place. It is therefore very likely that such an improvement of the truncation scheme compared to this work has considerable impact on the glueball spectrum.

Bibliography

[A+88] J. Ashman et al. A Measurement of the Spin Asymmetry and Determination of the Structure Function g(1) in Deep Inelastic Muon-Proton Scattering. *Phys.Lett.*, B206:364, 1988.

[A+09] C. Alexandrou et al. Low-lying baryon spectrum with two dynamical twisted mass fermions. *Phys.Rev.*, D80:114503, 2009.

[ADFM04] Reinhard Alkofer, W. Detmold, C.S. Fischer, and P. Maris. Analytic properties of the Landau gauge gluon and quark propagators. *Phys.Rev.*, D70:014014, 2004.

[AFLE05] Reinhard Alkofer, Christian S. Fischer, and Felipe J. Llanes-Estrada. Vertex functions and infrared fixed point in Landau gauge SU(N) Yang-Mills theory. *Phys. Lett.*, B611:279–288, 2005. [Erratum-ibid.670:460-461,2009].

[AFW08] Reinhard Alkofer, Christian S. Fischer, and Richard Williams. U(A)(1) anomaly and eta-prime mass from an infrared singular quark-gluon vertex. *Eur.Phys.J.*, A38:53–60, 2008.

[AHS09] Reinhard Alkofer, Markus Q. Huber, and Kai Schwenzer. Algorithmic derivation of Dyson-Schwinger Equations. *Comput.Phys.Commun.*, 180:965–976, 2009.

[AvS01] Reinhard Alkofer and Lorenz von Smekal. The Infrared behavior of QCD Green's functions: Confinement dynamical symmetry breaking, and hadrons as relativistic bound states. *Phys.Rept.*, 353:281, 2001.

[AWW02] Reinhard Alkofer, P. Watson, and H. Weigel. Mesons in a Poincare covariant Bethe-Salpeter approach. *Phys.Rev.*, D65:094026, 2002.

[B+93] G.S. Bali et al. A Comprehensive lattice study of SU(3) glueballs. *Phys.Lett.*, B309:378–384, 1993.

[BB09] Fabian Brau and Fabien Buisseret. Glueballs and statistical mechanics of the gluon plasma. *Phys.Rev.*, D79:114007, 2009.

[BC80] James S. Ball and Ting-Wai Chiu. ANALYTIC PROPERTIES OF THE VERTEX FUNCTION IN GAUGE THEORIES. 2. *Phys.Rev.*, D22:2550, 1980.

[Ber04] Juergen Berges. N-particle irreducible effective action techniques for gauge theories. *Phys.Rev.*, D70:105010, 2004.

[BGPP75] John F. Bolzan, Kevin A. Geer, William F. Palmer, and Stephen S. Pinsky. Zweig's Rule Violation, SU(3) Character of the Decay of the New Particles, and Pole Dominance Picture. *Phys.Lett.*, B59:351, 1975.

[BHL+07] Patrick O. Bowman, Urs M. Heller, Derek B. Leinweber, Maria B. Parappilly, Andre Sternbeck, et al. Scaling behavior and positivity violation of the gluon propagator in full QCD. *Phys.Rev.*, D76:094505, 2007.

[BLLY+08] Philippe Boucaud, J.P. Leroy, A. Le Yaouanc, J. Micheli, O. Pene, et al. On the IR behaviour of the Landau-gauge ghost propagator. *JHEP*, 0806:099, 2008.

[Blo01] Jacques C.R. Bloch. Multiplicative renormalizability of gluon and ghost propagators in QCD. *Phys.Rev.*, D64:116011, 2001.

[Blo03] Jacques C.R. Bloch. Two loop improved truncation of the ghost gluon Dyson-Schwinger equations: Multiplicatively renormalizable propagators and nonperturbative running coupling. *Few Body Syst.*, 33:111–152, 2003.

[BPP76] John F. Bolzan, William F. Palmer, and Stephen S. Pinsky. Forbidden Coupling and Inhibited Decay: A Study of Disconnection. *Phys.Rev.*, D14:3202, 1976.

[BRS76] C. Becchi, A. Rouet, and R. Stora. Renormalization of Gauge Theories. *Annals Phys.*, 98:287–321, 1976.

[BTB+10] A. Bazavov, D. Toussaint, C. Bernard, J. Laiho, C. DeTar, et al. Nonperturbative QCD simulations with 2+1 flavors of improved staggered quarks. *Rev.Mod.Phys.*, 82:1349–1417, 2010.

[Bui07] Fabien Buisseret. Meson and glueball spectra with the relativistic flux tube model. *Phys.Rev.*, C76:025206, 2007.

[Bui10] Fabien Buisseret. Glueballs, gluon condensate, and pure glue QCD below T(c). *Eur.Phys.J.*, C68:473–478, 2010.

[BV81] I.A. Batalin and G.A. Vilkovisky. Gauge Algebra and Quantization. *Phys.Lett.*, B102:27–31, 1981.

[CAD+06] Y. Chen, A. Alexandru, S.J. Dong, Terrence Draper, I. Horvath, et al. Glueball spectrum and matrix elements on anisotropic lattices. *Phys.Rev.*, D73:014516, 2006.

[Cal70] Jr. Callan, Curtis G. Broken scale invariance in scalar field theory. *Phys.Rev.*, D2:1541–1547, 1970.

[CB21] J. Chadwick and E. S. Bieler. The collisions of alpha particles with hydrogen nuclei. *Phil.Mag.*, 42:923–940, 1921.

[CCF+81] Carl E. Carlson, J.Joseph Coyne, Paul M. Fishbane, Franz Gross, and Sydney Meshkov. GLUEBALLS AND ODDBALLS: THEIR EXPERIMENTAL SIGNATURE. *Phys.Lett.*, B99:353, 1981.

[CFM80] J.Joseph Coyne, Paul M. Fishbane, and Sydney Meshkov. GLUEBALLS: THEIR SPECTRA, PRODUCTION AND DECAY. *Phys.Lett.*, B91:259, 1980.

[CG68] C.G. Callan and David J. Gross. Crucial Test of a Theory of Currents. *Phys.Rev.Lett.*, 21:311–313, 1968.

[CG69] Jr. Callan, Curtis G. and David J. Gross. High-energy electroproduction and the constitution of the electric current. *Phys.Rev.Lett.*, 22:156–159, 1969.

[Che61] G. Chew. S-Matrix Theory. 1961.

[Chu02] S. U. Chung. Two-photon Amplitudes of Quarkonia in Helicity Fornalism II. *BNL Preprint (unpublished)*, 2002.

[CJJ+74] A. Chodos, R.L. Jaffe, K. Johnson, Charles B. Thorn, and V.F. Weisskopf. A New Extended Model of Hadrons. *Phys.Rev.*, D9:3471–3495, 1974.

[CJT74] John M. Cornwall, R. Jackiw, and E. Tomboulis. Effective Action for Composite Operators. *Phys.Rev.*, D10:2428–2445, 1974.

[CM08] Alain Connes and Matilde Marcolli. Noncommutative geometry, quantum fields and motives. *American Physic Society Colloquium Publications*, 55, 2008.

[CM09] V. Crede and C.A. Meyer. The Experimental Status of Glueballs. *Prog.Part.Nucl.Phys.*, 63:74–116, 2009.

[CM10] Attilio Cucchieri and Tereza Mendes. Landau-gauge propagators in Yang-Mills theories at beta = 0: Massive solution versus conformal scaling. *Phys.Rev.*, D81:016005, 2010.

[CMM08] Attilio Cucchieri, Axel Maas, and Tereza Mendes. Three-point vertices in Landau-gauge Yang-Mills theory. *Phys.Rev.*, D77:094510, 2008.

[CW79] Peter Craven and Grace Wahba. Smoothing noisy data with spline functions. *Num.Math.*, 31:377–403, 1979.

[DFF+08] S. Durr, Z. Fodor, J. Frison, C. Hoelbling, R. Hoffmann, et al. Ab-Initio Determination of Light Hadron Masses. *Science*, 322:1224–1227, 2008.

[DFKS99] L. Driesen, J. Fromm, J. Kuhrs, and M. Stingl. Extended iterative scheme for QCD: Three point vertices. *Eur.Phys.J.*, A4:381–400, 1999.

[DGS11] D. Dudal, M.S. Guimaraes, and S.P. Sorella. Glueball masses from an infrared moment problem and nonperturbative Landau gauge. *Phys.Rev.Lett.*, 106:062003, 2011.

[DS99] L. Driesen and M. Stingl. Extended iterative scheme for QCD: The Four gluon vertex. *Eur.Phys.J.*, A4:401–419, 1999.

[EAKN10] G. Eichmann, R. Alkofer, A. Krassnigg, and D. Nicmorus. Nucleon mass from a covariant three-quark Faddeev equation. *Phys.Rev.Lett.*, 104:201601, 2010.

[Ede52] R. J. Eden. Quantum Field Theory of Bound States. I. Bound States in Weak Interaction. *Proc. R. Soc. Lond. A*, 215:133–146, 1952.

[Eic11] Gernot Eichmann. Nucleon electromagnetic form factors from the covariant Faddeev equation. *Phys.Rev.*, D84:014014, 2011.

[FA03] Christian S. Fischer and Reinhard Alkofer. Nonperturbative propagators, running coupling and dynamical quark mass of Landau gauge QCD. *Phys.Rev.*, D67:094020, 2003.

[FGML73] H. Fritzsch, Murray Gell-Mann, and H. Leutwyler. Advantages of the Color Octet Gluon Picture. *Phys.Lett.*, B47:365–368, 1973. Introduces the term 'color'.

[Fis03] Christian S. Fischer. Nonperturbative propagators, running coupling and dynamical mass generation in ghost - anti-ghost symmetric gauges in QCD. 2003. Ph.D. Thesis.

[Fis06] Christian S. Fischer. Infrared properties of QCD from Dyson-Schwinger equations. *J.Phys.G*, G32:R253–R291, 2006.

[FM75] Harald Fritzsch and Peter Minkowski. Psi Resonances, Gluons and the Zweig Rule. *Nuovo Cim.*, A30:393, 1975.

[FMP09] Christian S. Fischer, Axel Maas, and Jan M. Pawlowski. On the infrared behavior of Landau gauge Yang-Mills theory. *Annals Phys.*, 324:2408–2437, 2009.

[FN75] Peter G.O. Freund and Yoichiro Nambu. Dynamics in the Zweig-Iizuka Rule and a New Vector Meson Below 2-GeV/c**2. *Phys.Rev.Lett.*, 34:1645, 1975.

[FNW09] Christian S. Fischer, Dominik Nickel, and Richard Williams. On Gribov's supercriticality picture of quark confinement. *Eur.Phys.J.*, C60:47–61, 2009.

[FP67] L.D. Faddeev and V.N. Popov. Feynman Diagrams for the Yang-Mills Field. *Phys.Lett.*, B25:29–30, 1967.

[FP07] Christian S. Fischer and Jan M. Pawlowski. Uniqueness of infrared asymptotics in Landau gauge Yang-Mills theory. *Phys.Rev.*, D75:025012, 2007.

[FP09] Christian S. Fischer and Jan M. Pawlowski. Uniqueness of infrared asymptotics in Landau gauge Yang-Mills theory II. *Phys.Rev.*, D80:025023, 2009.

[Fra61] John Francis. The QR Transformation: A Unitary Analogue to the LR Transformation—Part 1. *The Computer Journal*, 4(3):265–271, 1961.

[Fra62] John Francis. The QR Transformation—Part 2. *The Computer Journal*, 4(4):332–345, 1962.

[FW08] Christian S. Fischer and Richard Williams. Beyond the rainbow: Effects from pion back-coupling. *Phys.Rev.*, D78:074006, 2008.

[FW09] Christian S. Fischer and Richard Williams. Probing the gluon self-interaction in light mesons. *Phys.Rev.Lett.*, 103:122001, 2009.

[GLS69] David J. Gross and Chris H. Llewellyn Smith. High-energy neutrino - nucleon scattering, current algebra and partons. *Nucl.Phys.*, B14:337–347, 1969.

[GM09] H. Geiger and E. Marsden. On a Diffuse Reflection of the alpha-Particles. *Proc. Roy. Soc. A*, 82:495–500, 1909.

[GM64] Murray Gell-Mann. A Schematic Model of Baryons and Mesons. *Phys.Lett.*, 8:214–215, 1964.

[GML51] M Gell-Mann and F Low. Bound states in quantum field theory. *Physical Review*, 84(2):350–354, 1951.

[GMN64] Murray Gell-Mann and Yuval Neemam. The Eightfold way: a review with a collection of reprints. 1964. W. A. Benjamin, Publisher (QCD161:G4).

[GR80] Israil Solomonowitsch Gradshteyn and Josif Moissejewitsch Ryzhik. Table of Integrals, Series and Products; 2nd Revised edition. *Academic Press Inc.*, 1980.

[Gre64] O.W. Greenberg. Spin and Unitary Spin Independence in a Paraquark Model of Baryons and Mesons. *Phys.Rev.Lett.*, 13:598–602, 1964. Introduces the symmetry that was to become color.

[Gri61] V.N. Gribov. Asymptotic behavior of the scattering amplitude at high-energies. *Nucl.Phys.*, 22:249, 1961.

[Gri78] V.N. Gribov. Quantization of Nonabelian Gauge Theories. *Nucl.Phys.*, B139:1, 1978.

[Gro05] D. J. Gross. The discovery of asymptotic freedom and the emergence of QCD. *Proc. Nat. Acad. Sci.*, 102:9099–9108, 2005. [Int.J.Mod.Phys.A20:5717-5740,2005].

[GSG⁺08] Peng Guo, Adam P. Szczepaniak, Giuseppe Galata, Andrea Vassallo, and Elena Santopinto. Gluelump spectrum from Coulomb gauge QCD. *Phys.Rev.*, D77:056005, 2008.

[GW73a] D.J. Gross and Frank Wilczek. Asymptotically Free Gauge Theories. 1. *Phys.Rev.*, D8:3633–3652, 1973.

[GW73b] D.J. Gross and Frank Wilczek. Ultraviolet Behavior of Nonabelian Gauge Theories. *Phys.Rev.Lett.*, 30:1343–1346, 1973.

[GW74] D.J. Gross and Frank Wilczek. ASYMPTOTICALLY FREE GAUGE THEORIES. 2. *Phys.Rev.*, D9:980–993, 1974.

[Hac95] Wolfgang Hackbusch. Integral Equations: Theory and Numerical Treatment. *Birkhauser Basel*, 1995.

[HAS10] Markus Q. Huber, Reinhard Alkofer, and Silvio P. Sorella. Infrared analysis of Dyson-Schwinger equations taking into account the Gribov horizon in Landau gauge. *Phys.Rev.*, D81:065003, 2010.

[HMMP06] A. Hart, C. McNeile, Christopher Michael, and J. Pickavance. A lattice study of the masses of singlet 0++ mesons. *Phys. Rev.*, D74:114504, 2006.

[HN65] M.Y. Han and Yoichiro Nambu. Three Triplet Model with Double SU(3) Symmetry. *Phys.Rev.*, 139:B1006–B1010, 1965.

[HSA10] Markus Q. Huber, Kai Schwenzer, and Reinhard Alkofer. On the infrared scaling solution of SU(N) Yang-Mills theories in the maximally Abelian gauge. *Eur.Phys.J.*, C68:581–600, 2010.

[HvSA98] Andreas Hauck, Lorenz von Smekal, and Reinhard Alkofer. Solving the gluon Dyson-Schwinger equations in the Mandelstam approximation. *Comput.Phys.Commun.*, 112:149, 1998.

[IMPS+07] E.-M. Ilgenfritz, M. Muller-Preussker, A. Sternbeck, A. Schiller, and I.L. Bogolubsky. Landau gauge gluon and ghost propagators from lattice QCD. *Braz.J.Phys.*, 37:193–200, 2007.

[IOS66] Jugoro Iizuka, Kunihiko Okada, and Okiyasu Shito. Systematics and phenomenology of boson mass levels. 3. *Prog.Theor.Phys.*, 35:1061–1073, 1966.

[JJ76] R.L. Jaffe and K. Johnson. Unconventional States of Confined Quarks and Gluons. *Phys.Lett.*, B60:201, 1976.

[Joh10] Tord Johansson. Hadron physics with PANDA. *PoS*, BORMIO2010:031, 2010.

[JW59] M. Jacob and G.C. Wick. On the general theory of collisions for particles with spin. *Annals Phys.*, 7:404–428, 1959.

[KePeSe65] B.N. Kursunoglu (ed.), A. Perlmutter (ed.), and I. Sanmar (ed.). Proceedings of the 2nd Coral Gables Conference on Symmetry Principles at High Energy. pages 274–285, 1965.

[KF08] Christian Kellermann and Christian S. Fischer. The Running coupling from the four-gluon vertex in Landau gauge Yang-Mills theory. *Phys.Rev.*, D78:025015, 2008.

[KO79] Taichiro Kugo and Izumi Ojima. Local Covariant Operator Formalism of Nonabelian Gauge Theories and Quark Confinement Problem. *Prog.Theor.Phys.Suppl.*, 66:1, 1979.

[Kon09] Kei-Ichi Kondo. Decoupling and scaling solutions in Yang-Mills theory with the Gribov horizon. 2009.

[KS00] A.B. Kaidalov and Yu.A. Simonov. Glueball masses and Pomeron trajectory in nonperturbative QCD approach. *Phys.Lett.*, B477:163–170, 2000.

[KSS76] John B. Kogut, Donald K. Sinclair, and Leonard Susskind. A Quantitative Approach to Low-Energy Quantum Chromodynamics. *Nucl.Phys.*, B114:199, 1976.

[KUB+96] L.P. Kaptari, A.Yu. Umnikov, S.G. Bondarenko, K.Yu. Kazakov, F.C. Khanna, et al. Bethe-Salpeter amplitudes and static properties of the deuteron. *Phys.Rev.*, C54:986–1005, 1996.

[Kug95] Taichiro Kugo. The Universal renormalization factors Z(1) / Z(3) and color confinement condition in nonAbelian gauge theory. pages 107–119, 1995.

[KZ07] Eberhard Klempt and Alexander Zaitsev. Glueballs, Hybrids, Multiquarks. Experimental facts versus QCD inspired concepts. *Phys.Rept.*, 454:1–202, 2007.

[Lan48] L. D. Landau. The moment of a 2-photon systemy. *Dokl. Akad. Nauk*, 60:207, 1948.

[Lan09] Jens Soren Lange. The PANDA experiment: Hadron physics with antiprotons at FAIR. *Int.J.Mod.Phys.*, A24:369–376, 2009.

[LEBC06] Felipe J. Llanes-Estrada, Pedro Bicudo, and Stephen R. Cotanch. Oddballs and a low odderon intercept. *Phys.Rev.Lett.*, 96:081601, 2006.

[LECdAB+02] Felipe J. Llanes-Estrada, Stephen R. Cotanch, Pedro J. de A. Bicudo, J. Emilio F.T. Ribeiro, and Adam P. Szczepaniak. QCD glueball Regge trajectories and the Pomeron. *Nucl.Phys.*, A710:45–54, 2002.

[LP55] L.D. Landau and I.Ya. Pomeranchuk. On Point interactions in quantum electrodynamics. *Dokl.Akad.Nauk Ser.Fiz.*, 102:489, 1955. Also published in Collected Papers of L.D. Landua. Edited by D. Ter Haar. Pergamon Press, 1965. pp. 654-658.

[LS69] C.H. Llewellyn-Smith. A relativistic formulation for the quark model for mesons. *Annals Phys.*, 53:521–558, 1969.

[LvS02] Christoph Lerche and Lorenz von Smekal. On the infrared exponent for gluon and ghost propagation in Landau gauge QCD. *Phys.Rev.*, D65:125006, 2002.

[LW02] Da Qing Liu and Ji Min Wu. The First calculation for the mass of the ground 4++ glueball state on lattice. *Mod.Phys.Lett.*, A17:1419–1430, 2002.

[LWC02] Da-Qing Liu, Ji-Min Wu, and Ying Chen. A New approach to construct the operator on lattice for the calculation of glueball masses. *High Energy Phys.Nucl.Phys.*, 26:222–229, 2002.

[Maa10a] Axel Maas. Constructing non-perturbative gauges using correlation functions. *Phys.Lett.*, B689:107–111, 2010.

[Maa10b] Axel Maas. On gauge fixing. *PoS*, LATTICE2010:279, 2010.

[Man55] S. Mandelstam. Dynamical Variables in the Bethe-Salpeter Formalism. *Proceedings of the Royal Society A: Mathematical, Physical and Engineering Sciences*, 233(1193):248–266, December 1955.

[Mar02] P. Maris. Electromagnetic, weak, and strong interactions of light mesons. *PiN Newslett.*, 16:213–218, 2002.

[McN09] Craig McNeile. Lattice status of gluonia/glueballs. *Nucl.Phys.Proc.Suppl.*, 186:264–267, 2009.

[MKV09] Vincent Mathieu, Nikolai Kochelev, and Vicente Vento. The Physics of Glueballs. *Int.J.Mod.Phys.*, E18:1–49, 2009.

[MP78] William J. Marciano and H. Pagels. Quantum Chromodynamics. 1978.

[MP99] Colin J. Morningstar and Mike J. Peardon. The Glueball spectrum from an anisotropic lattice study. *Phys.Rev.*, D60:034509, 1999.

[MR03] Pieter Maris and Craig D. Roberts. Dyson-Schwinger equations: A Tool for hadron physics. *Int.J.Mod.Phys.*, E12:297–365, 2003.

[MSSB08] Vincent Mathieu, Claude Semay, and Bernard Silvestre-Brac. Semirelativistic potential model for three-gluon glueballs. *Phys.Rev.*, D77:094009, 2008.

[MT00] Pieter Maris and Peter C. Tandy. The pi, K+, and K0 electromagnetic form-factors. *Phys.Rev.*, C62:055204, 2000.

[MT05] Harvey B. Meyer and Michael J. Teper. Glueball Regge trajectories and the pomeron: A Lattice study. *Phys.Lett.*, B605:344–354, 2005.

[N$^+$10] K. Nakamura et al. Review of particle physics. *J.Phys.G*, G37:075021, 2010.

[Nai05] V.P. Nair. Quantum field theory: A modern perspective. 2005.

[Nak69] N. Nakanishi. A GENERAL SURVEY OF THE THEORY OF THE BETHE–SALPETER EQUATION. 1969.

[Nam50] Yoichiro Nambu. The use of the Proper Time in Quantum Electrodynamics. *Prog.Theor.Phys.*, 5:82–94, 1950. In *Eguchi, T. (ed.), Nishijima, K. (ed.): Broken symmetry*.

[NJL61] Yoichiro Nambu and G. Jona-Lasinio. Dynamical Model of Elementary Particles Based on an Analogy with Superconductivity. 1. *Phys.Rev.*, 122:345–358, 1961.

[NRW01] Ferenc Niedermayer, Philipp Rufenacht, and Urs Wenger. Fixed point gauge actions with fat links: Scaling and glueballs. *Nucl.Phys.*, B597:413–450, 2001.

[Oku63] S. Okubo. Phi meson and unitary symmetry model. *Phys.Lett.*, 5:165–168, 1963.

[OS73] Konrad Osterwalder and Robert Schrader. AXIOMS FOR EUCLIDEAN GREEN'S FUNCTIONS. *Commun.Math.Phys.*, 31:83–112, 1973.

[OS75] Konrad Osterwalder and Robert Schrader. Axioms for Euclidean Green's Functions. 2. *Commun.Math.Phys.*, 42:281, 1975.

[OS09] O. Oliveira and P.J. Silva. Does The Lattice Zero Momentum Gluon Propagator for Pure Gauge SU(3) Yang-Mills Theory Vanish in the Infinite Volume Limit? *Phys.Rev.*, D79:031501, 2009.

[OSIS07] O. Oliveira, P.J. Silva, E.M. Ilgenfritz, and A. Sternbeck. The Gluon propagator from large asymmetric lattices. *PoS*, LAT2007:323, 2007.

[Pau55] W. Pauli. Niels Bohr and the Development of Physics. pages 52–69, 1955.

[Pet07] K. Peters. Charmonium and exotic hadrons at PANDA. *Int.J.Mod.Phys.*, E16:919–924, 2007.

[PLNvS04] Jan M. Pawlowski, Daniel F. Litim, Sergei Nedelko, and Lorenz von Smekal. Infrared behavior and fixed points in Landau gauge QCD. *Phys.Rev.Lett.*, 93:152002, 2004.

[Pom58] I. Y. Pomeranchuk. Equality between the Interaction Cross Sections of High Energy Nucleons and Antinucleons . *Zh.Eksp. i Teor Fiz*, 34:725, 1958.

[RIGM10] Christopher M. Richards, Alan C. Irving, Eric B. Gregory, and Craig McNeile. Glueball mass measurements from improved staggered fermion simulations. *Phys.Rev.*, D82:034501, 2010.

[RSS+99] D.G. Robertson, E.S. Swanson, A.P. Szczepaniak, C.R. Ji, and S.R. Cotanch. Renormalized effective QCD Hamiltonian: Gluonic sector. *Phys.Rev.*, D59:074019, 1999.

[Rut11] E. Rutherford. The scattering of alpha and beta particles by matter and the structure of the atom. *Phil.Mag.*, 21:669–688, 1911.

[Rut19] E. Rutherford. Collision of Particles with Light Atoms IV. An Anomalous Effect in Nitrogen. *Phil.Mag.*, 37:581–587, 1919.

[RW94] Craig D. Roberts and Anthony G. Williams. Dyson-Schwinger equations and their application to hadronic physics. *Prog.Part.Nucl.Phys.*, 33:477–575, 1994.

[SAEVCA11] Helios Sanchis-Alepuz, Gernot Eichmann, Selym Villalba-Chavez, and Reinhard Alkofer. Delta and Omega masses in a three-quark covariant Faddeev approach. *Phys.Rev.*, D84:096003, 2011. 9 pages, 4 figures, 3 tables.

[SB51] EE Salpeter and HA Bethe. A relativistic equation for bound-state problems. *Physical Review*, 84(6):1232–1242, 1951.

[Sch51a] J. Schwinger. On the Green's functions of quantized fields. I. *Proceedings of the National Academy of Sciences*, 37(7):451–455, 1951.

[Sch51b] J. Schwinger. On the Green's functions of quntized fields. II. *Proceedings of the National Academy of Sciences*, 37(7):455–459, 1951.

[Sch08] W. Schleifenbaum. Nonperturbative aspects of Yang-Mills theory. 2008.

[She09] M.R. Shepherd. The GLUEX experiment. *AIP Conf.Proc.*, 1182:816–819, 2009.

[SIMP+06] A. Sternbeck, E.-M. Ilgenfritz, M. Muller-Preussker, A. Schiller, and I.L. Bogolubsky. Lattice study of the infrared behavior of QCD Green's functions in Landau gauge. *PoS*, LAT2006:076, 2006.

[SMWA05] Wolfgang Schleifenbaum, Axel Maas, Jochen Wambach, and Reinhard Alkofer. Infrared behaviour of the ghost-gluon vertex in Landau gauge Yang-Mills theory. *Phys.Rev.*, D72:014017, 2005.

[SS03] Adam P. Szczepaniak and Eric S. Swanson. The Low lying glueball spectrum. *Phys.Lett.*, B577:61–66, 2003.

[Sto64] R. Stoops. The Quantum Theory of Fields: The 12th Solvay Conference. Proceedings of the 12th Solvay Conference on Physics. Brussels, Belgium. October 1962. 1964.

[Sym70] K. Symanzik. Small distance behavior in field theory and power counting. *Commun.Math.Phys.*, 18:227–246, 1970.

[Szc05] A.P. Szczepaniak. Gluonic excitations and the GlueX experiment. *J.Phys.Conf.Ser.*, 9:315–320, 2005.

[Tay71] J.C. Taylor. Ward Identities and Charge Renormalization of the Yang-Mills Field. *Nucl.Phys.*, B33:436–444, 1971.

[Tri85] Francesco Giacomo Tricomi. Integral Equations. *Dover Publications*, 1985.

[Tyu75] I.V. Tyutin. Gauge Invariance in Field Theory and Statistical Physics in Operator Formalism. 1975.

[vB97] Pierre van Baal. Gribov ambiguities and the fundamental domain. 1997.

[vBRM+86] E. van Beveren, T.A. Rijken, K. Metzger, C. Dullemond, G. Rupp, et al. A Low Lying Scalar Meson Nonet in a Unitarized Meson Model. *Z.Phys.*, C30:615–620, 1986.

[vHedSeFe65] L. van Hove (ed.), A. de Shalit (ed.), and H. Feshbach (ed.). Preludes in Theoretical Physics in Honor of V. F. Weisskopf. pages 133–1425, 1965.

[vSHA98] Lorenz von Smekal, Andreas Hauck, and Reinhard Alkofer. A solution to coupled Dyson-Schwinger equations for gluons and ghosts in Landau gauge. *Annals Phys.*, 267:1, 1998.

[WC04] Peter Watson and Wolfgang Cassing. Unquenching the quark-antiquark Green's function. *Few Body Syst.*, 35:99–115, 2004.

[Wei60] Steven Weinberg. High-energy behavior in quantum field theory. *Phys.Rev.*, 118:838–849, 1960.

[Wei64a] Steven Weinberg. Feynman Rules for Any Spin. *Phys.Rev.*, 133:B1318–B1332, 1964.

[Wei64b] Steven Weinberg. Feynman Rules for Any Spin. 2. Massless Particles. *Phys.Rev.*, 134:B882–B896, 1964.

[Wei95] Steven Weinberg. The quantum theory of fields. Vol. 1: Foundations. 1995.

[Wei96] Steven Weinberg. The quantum theory of fields. Vol. 2: Modern applications. 1996.

[Wie44] Helmut Wielandt. Beiträge zur mathematischen Behandlung komplexer Eigenwertprobleme, Teil V: Bestimmung höherer Eigenwerte durch gebrochene Iteration. *Bericht B 44/J/37, Aerodynamische Versuchsanstalt Göttingen*, 1944.

[WR08] Peter Watson and Hugo Reinhardt. Two-point functions of Coulomb gauge Yang-Mills theory. *Phys.Rev.*, D77:025030, 2008.

[Yan50] Chen-Ning Yang. Selection Rules for the Dematerialization of a Particle Into Two Photons. *Phys.Rev.*, 77:242–245, 1950.

[Zem65] Charles Zemach. Use of angular momentum tensors. *Phys.Rev.*, 140:B97–B108, 1965.

[Zih10] B. Zihlmann. GlueX a new facility to search for gluonic degrees of freedom in mesons. *AIP Conf.Proc.*, 1257:116–120, 2010.

[Zwa82] Daniel Zwanziger. NONPERTURBATIVE MODIFICATION OF THE FADDEEV-POPOV FORMULA AND BANISHMENT OF THE NAIVE VACUUM. *Nucl.Phys.*, B209:336, 1982.

[Zwa89] D. Zwanziger. LOCAL AND RENORMALIZABLE ACTION FROM THE GRIBOV HORIZON. *Nucl. Phys.*, B323:513–544, 1989.

[Zwe64] G. Zweig. AN SU(3) MODEL FOR STRONG INTERACTION SYMMETRY AND ITS BREAKING. 2. 1964. Published in 'Developments in the Quark Theory of Hadrons'. Volume 1. Edited by D. Lichtenberg and S. Rosen. Nonantum, Mass., Hadronic Press, 1980. pp. 22-101.

Acknowledgements

There have been several people I would never have been able to finish this thesis without and I would like to express my gratitude.

First I thank Christian Fischer for giving me the opportunity for this thesis, for his efforts to extend the time to finish it and for his supervision. Also for giving me the opportunity to participate in the summer-school in Saalburg.

Richard Williams deserves my gratitude for his support, enlightening discussions, cross-checks and lots of annoying yet legitimate questions.

Also I thank Lorenz von Smekal for helpful discussions about invariant amplitudes, without which the corresponding section of this thesis probably never would have been completed.

Furthermore I thank the whole lot of people in the theory group of the institute for nuclear physics, especially the "Kinderzimmer" and all those who belong to it in their hearts for a really good time.

Very importantly I thank my aunt and uncle Martina and Ronald Schwenke for their encouragement and unconditional support, which saved me from a plethora of distractions.

And finally I want to thank Sarah-Mailin Langhagen for here support, encouragement, care and love, keeping my spirits up, when things turned out for the worst again.

This work was supported by the Helmholtz-University Young Investigator Grant No. VH-NG-332 and by the Helmholtz International Center for FAIR within the LOEWE program.

Lebenslauf

Name	Christian Kellermann
Geburtsdatum	08.09.1979
Geburtsort	Halle/Saale
Staatsangehörigkeit	deutsch
Familienstand	ledig

Bildungsweg

1986 - 1990	Besuch der Aueschule (Grundschule) in Dietzenbach
1990 - 1999	Besuch des Gymnasiums an der Heinrich-Mann Schule in Dietzenbach.
1999	Abschluß: Abitur; Gesamtnote: 1,9.
1999 - 2000	Zivildienst
2000 - 2006	Studium der Physik an der Technischen Universität Darmstadt
2006 - 2007	Anfertigung der Diplomarbeit am Institut für Kernphysik der Technischen Universität Darmstadt. Titel: „*The four-gluon vertex in Landau gauge Yang-Mills theory*“.
2007	Abschluß mit Grad Diplom-Physiker. Gesamtnote: Sehr Gut.
seit 2007	Promotionsstudium und Anfertigung der Dissertation am Institut für Kernphysik der Technischen Universität Darmstadt.
2007 - 2010	Wissenschaftlicher Mitarbeiter am Institut für Kernphysik der Technischen Universität Darmstadt.
2010 - 2011	Stipendiat des Helmholtz International Center for FAIR (HIC for FAIR)

Erklärung:

Ich versichere, dass ich diese Arbeit selbständig verfasst und keine anderen als die angegebenen Quellen und Hilfsmittel benutzt habe.
Vor dem Einreichen dieser Dissertation habe ich keinen anderen Dissertationsversuch unternommen.

Darmstadt, den 14. Oktober 2011

....
Unterschrift